NATIONAL
ACADEMIES
PRESS
Washington, DC

Forecasting the Ocean

The 2025–2035 Decade of Ocean Science

Committee for the 2025–2035
Decadal Survey of Ocean Sciences
for the National Science Foundation

Ocean Studies Board

Division on Earth and Life Studies

Consensus Study Report

NATIONAL ACADEMIES PRESS 500 Fifth Street, NW Washington, DC 20001

This activity was supported by a contract between the National Academy of Sciences and the National Science Foundation (Contract No. AWD-001795). Any opinions, findings, conclusions, or recommendations expressed in this publication do not necessarily reflect the views of any organization or agency that provided support for the project.

International Standard Book Number-13: 978-0-309-72222-3
International Standard Book Number-10: 0-309-72222-5
Digital Object Identifier: https://doi.org/10.17226/27846

This publication is available from the National Academies Press, 500 Fifth Street, NW, Keck 360, Washington, DC 20001; (800) 624-6242 or (202) 334-3313; http://www.nap.edu.

Printed in the United States of America.

Suggested citation: National Academies of Sciences, Engineering, and Medicine. 2025. *Forecasting the Ocean: The 2025–2035 Decade of Ocean Science*. Washington, DC: The National Academies Press. https://doi.org/10.17226/27846.

The **National Academy of Sciences** was established in 1863 by an Act of Congress, signed by President Lincoln, as a private, nongovernmental institution to advise the nation on issues related to science and technology. Members are elected by their peers for outstanding contributions to research. Dr. Marcia McNutt is president.

The **National Academy of Engineering** was established in 1964 under the charter of the National Academy of Sciences to bring the practices of engineering to advising the nation. Members are elected by their peers for extraordinary contributions to engineering. Dr. John L. Anderson is president.

The **National Academy of Medicine** (formerly the Institute of Medicine) was established in 1970 under the charter of the National Academy of Sciences to advise the nation on medical and health issues. Members are elected by their peers for distinguished contributions to medicine and health. Dr. Victor J. Dzau is president.

The three Academies work together as the **National Academies of Sciences, Engineering, and Medicine** to provide independent, objective analysis and advice to the nation and conduct other activities to solve complex problems and inform public policy decisions. The National Academies also encourage education and research, recognize outstanding contributions to knowledge, and increase public understanding in matters of science, engineering, and medicine.

Learn more about the National Academies of Sciences, Engineering, and Medicine at **www.nationalacademies.org**.

Reviewers

This consensus study report was reviewed in draft form by individuals chosen for their diverse perspectives and technical expertise. The purpose of this independent review is to provide candid and critical comments that will assist the National Academies of Sciences, Engineering, and Medicine in making each published report as sound as possible and to ensure that it meets the institutional standards for quality, objectivity, evidence, and responsiveness to the study charge. The review comments and draft manuscript remain confidential to protect the integrity of the deliberative process.

We thank the following individuals for their review of this report:

MARK ABBOTT, Woods Hole Oceanographic Institution
ROSIE 'ANOLANI ALEGADO, University of Hawai'i at Mānoa
PATRICK CHRISTIE, University of Washington
SCOTT DONEY, University of Virginia
SARAH T. GILLE, Scripps Institution of Oceanography
PATRICK HEIMBACH, University of Texas
KENNETH JOHNSON, Monterey Bay Aquarium Research Institute
BO BARKER JØRGENSEN, Arhaus University
LISA LEVIN, Scripps Institution of Oceanography
CRAIG MCLEAN, National Oceanic and Atmospheric Administration (*retired*)
STEVE MURAWSKI, University of South Florida
JOHN ORCUTT, Scripps Institution of Oceanography
AMELIA SHEVENELL, University of South Florida
AMY TRICE, Northeast Regional Ocean Council
MARTIN VISBECK, GEOMAR
EARLE WILSON, Stanford University

Although the reviewers listed above provided many constructive comments and suggestions, they were not asked to endorse the conclusions or recommendations of this report, nor did they see the final draft before its release. The review of this report was overseen by **ANDREW SOLOW,** Woods Hole Oceanographic Institution, and **DAVID KARL (NAS),** University of Hawai'i at Mānoa. They were responsible for making certain that an independent examination of this report was carried out in accordance with the standards of the National Academies and that all review comments were carefully considered. Responsibility for the final content rests entirely with the authoring committee and the National Academies.

Acknowledgments

The committee thanks the following individuals for their contributions during the study process, especially for enriching and informing the discussions at the open session meetings of the committee: Merryl Alber (University of Georgia), Rosie Alegado (University of Hawai'i), James Allen (National Science Foundation [NSF]), Katie Arkema (Pacific Northwest National Laboratory), Katherine Barbeau (University of California, San Diego), Paul Barber (University of California, Los Angeles), Jack Barth (Oregon State University/Northwest Association of Networked Ocean Observing Systems [NANOOS]), Joey Bernhardt (University of Guelph), Jennifer Biddle (University of Delaware), Donna Blackman (University of California, San Diego), Tim Boyer (National Oceanic and Atmospheric Administration [NOAA], National Centers for Environmental Information [NCEI]), Stefanie Brachfeld (Montclair State University), Carl Brenner (U.S. Science Support Program), Emily Brodsky (University of California, Santa Cruz), Deborah Bronk (Bigelow Laboratory for Ocean Sciences/University-National Oceanographic Laboratory System [UNOLS] Chair), Almesha Campbell (Jackson State University), Gabrielle Canonico (NOAA), Jackie Caplan-Auerbach (Western Washington University), Eric Chassignet (Florida State University), Shuyi Chen (University of Washington), Eric Cordes (Temple University), Sarah Davies (Boston University), John Delaney (University of Washington), Ed Dever (Oregon State University), Steven D'Hondt (University of Rhode Island), Emmett Duffy (Smithsonian Institute), Rose Dufour (NSF), Sonya Dyhrman (Lamont-Doherty Earth Observatory), Jim Edson (Woods Hole Oceanographic Institution), Duane Elgin (author of *Choosing Earth*), Melanie Fewings (Oregon State University), Patrick Fulton (Cornell University), Amy Gartman (U.S. Geological Survey), Corey Garza (University of Washington), Peter Gerstoft (Scripps Institution of Oceanography), Sarah Giddings (Scripps Institution of Oceanography), Sarah Gille (Scripps Institution of Oceanography), Holly Greening (Tampa Bay Estuary Program), Karen Grissom (NOAA, NCEI), Zachary Gold (NOAA), Sean Gulick (The University of Texas at Austin), Patrick Heimbach (The University of Texas at Austin), David Hodell (University of Cambridge), Russ Hopcroft (University of Alaska Fairbanks), Bruce Howe (University of Hawai'i at Mānoa), Celli Hull (Yale University), David Hutchins (University of Southern California), Minoru Ikehara (Kochi University), Fumio Inagaki (Japanese Agency for Marine-Earth Science and Technology), Alex Isern (NSF), Ken Johnson (Monterey Bay Aquarium Research Institute), Kevin Johnson (NSF), Brandon Jones (NSF), Henry Jones (University of Southern Mississippi), Maria Kavanaugh (Oregon State University), Deborah Kelley (University of Washington), Brandi Kiel Reese (University of South Alabama), Frieder Klein (Woods Hole Oceanographic Institution), Anthony Koppers (Oregon State University), David Koweek (OceanVisions), Larry Krissek (The Ohio State University), Kristy Kroeker (University of California, Santa Cruz), Jessica Labonté (Texas A&M University at Galveston), Adriane Lam (Binghamton University), Craig Lee (University of Washington), Chris Lowery (The University of Texas at Austin), Kelly Lucas (University of Southern Mississippi), Mitch Malone (Texas A&M University), Kathie Marsaglia (California State University, Northridge), Robert McKay (Victoria University of Wellington), Galen McKinley (Columbia University), Margaret McManus (University of Hawai'i), Lisa McNeil (Southampton University), Diego Melgar (University of Oregon), Charna Meth (International Ocean Discovery Program Science Support Office), Steve Murawski (University of South Florida), Mark Ohman (University of California, San Diego), Kirsten Oleson (Pacific Research on Island Solutions for Adaptation), Beth Orcutt (Bigelow Laboratory for Ocean Sciences), Heiko Pälike (University of Bremen), Ross Parnell-Turner (Scripps Intuition of Oceanography), Malin Pinsky (University of California, Santa Cruz), Charlie Plybon (Surfrider Foundation), Julie Pullen (Propeller Ventures), Kanna Rajan (RAND), Becky Robinson (University of Rhode Island), Yair Rosenthal (Rutgers University), Doug Russell (UNOLS), Demian Saffer (The University of Texas at Austin), Prasanna Sattigeri (International Business Machines [IBM]), Robert Shearman (Office of Naval Research), Joe Schumacker (NANOOS), Daniel Sigman (Princeton University), David Smith (University of Rhode Island), Heidi Sosik (Woods Hole Oceanographic Institution), Robert Sparrock (Office of Naval Research),

Robert Sterner (University of Minnesota Duluth), Mike Stukel (Florida State University), Chijun Sun (National Center for Atmospheric Research), Jason Sylvan (Texas A&M University), Lynne Talley (Scripps Institution of Oceanography), Allyson Tessin (Kent State University), Jeremy Testa (University of Maryland Center for Environmental Science Chesapeake Biological Laboratory), LuAnne Thompson (University of Washington), Masako Tominaga (Woods Hole Oceanographic Institution), Aradhna Tripati (University of California, Los Angeles), Robert Twilley (Louisiana State University), Alexis Valauri-Orton (The Ocean Foundation), Maureen Walczak (Oregon State University), Allen Walker (NSF's Technology, Innovation, and Partnership Directorate), Shelby Walker (NSF), Jessica Warren (University of Delaware), Doug Wiens (Washington University), Susan Wijffels (Woods Hole Oceanographic Institution), William Wilcock (University of Washington), Trevor Williams (Texas A&M University), Warren Wood (U.S. Naval Research Laboratory), Christine Yifeng Chen (Lawrence Livermore National Laboratory), and Zhongwen Zhan (California Institute of Technology). Their input was critical to the completion of the committee's work.

The committee would also like to thank our primary contact at NSF's Division of Ocean Sciences, Jim McManus, for his efforts in developing and sponsoring this study and for providing important documents and support upon the committee's request.

Contents

Boxes, Figures, and Tables

BOXES

FIGURES

TABLES

Preface

Our committee members represent a broad range of backgrounds and shared many different perspectives on the content of our report. We thank them for their hard work, for the respect they showed each other during our many in-person and online discussions, and for their contributions. It was an easy group to coordinate. We also thank the presenters and those who responded to our requests for information. Much of their input is represented in the report. We thank the staff of the National Academies of Sciences, Engineering, and Medicine, who did such an excellent job arranging the logistics for our meetings; we are especially grateful to Kelly Oskvig for her guidance throughout the study and her many contributions to the organization and writing of the report. She provided the glue that held us together throughout the study.

Our statement of task is significantly more complicated than the one that led to the 2015 report *Sea Change: 2015–2025 Decadal Survey of Ocean Sciences*, whose statement of task was focused on achieving a balance between funds spent on major facilities (academic fleet, ocean observatories, and ocean drilling) and those supporting ocean research programs. In contrast, the statement of task for the present report requested the committee's thoughts on future research directions and the tools required to support that research. In recognition of the changes to the research environment that have occurred during the past decade, we were also asked for recommendations on innovative research strategies and workforce training. The latter is particularly important given the significant demographic changes occurring in our country, as well as the recognition that we need to pay more attention to different sources and forms of knowledge. We not only have to engage a broad demographic to provide scientists and technicians to work in our field, but we also have to demonstrate to the U.S. population that it is getting value from the ocean research that its tax dollars support. Thus our committee appropriately spent as much time discussing and then writing about research strategies and workforce as on research questions and infrastructure.

We hope readers will appreciate and support our recommendations for future research directions, the tools that we believe are required, the different approach to research that we believe is necessary, and the efforts that are required to recruit a future ocean sciences workforce. We have much work to do, and we hope that everyone in the ocean sciences community will bring their talents to realizing this vision. We need all hands on deck.

Tuba Özkan-Haller, *Co-Chair*
James Yoder, *Co-Chair*
Committee on the 2025–2035 Decadal Survey of
Ocean Sciences for the National Science Foundation

Acronyms and Abbreviations

AI	artificial intelligence
AMOC	Atlantic meridional overturning circulation
ARF	Academic Research Fleet
ASV	autonomous surface vehicle
AUV	autonomous underwater vehicle
BCP	biological carbon pump
BGC-Argo	biogeochemical Argo
BOEM	Bureau of Ocean Energy Management
CARE	collective benefit, authority to control, responsibility, ethics
CISE	Computer and Information Science and Engineering (Directorate)
CO$_2$	carbon dioxide
CoPe	Coastlines and People
DOE	Department of Energy
FAIR	findable, accessible, interoperable, and reusable
GO-SHIP	Global Ocean Ship-based Hydrographic Investigations Program
HAB	harmful algal bloom
HOV	human-occupied vehicle
IODP	International Ocean Discovery Program
IPCC	Intergovernmental Panel on Climate Change
LDEO	Lamont-Doherty Earth Observatory
LEAP	Legacy Asset Project
mCDR	marine carbon dioxide removal
NAML	National Association of Marine Laboratories
NAVO	Naval Oceanographic Office
NDSF	National Deep Submergence Facility
NOAA	National Oceanic and Atmospheric Administration
NOPP	National Ocean Partnership Program
NSF	National Science Foundation
OBSIC	Ocean Bottom Seismic Instrument Center
OCE	Division of Ocean Sciences
ODZ	oxygen-deficient zone
ONR	Office of Natural Resources
OOI	Ocean Observatories Initiative
PEER	Partnerships for Enhanced Engagement in Research

RCR/V   regional-class research vessel
RISE   Research, Innovation, Synergies and Education
ROV   remotely operated vehicle
R/V   research vessel

SBIR   small business innovation research
SGIP   Seafloor Geodetic Instrument Pool
SMART   Science Monitoring and Reliable Telecommunications (cables)
STEM   science, technology, engineering, and mathematics
STTR   small business technology transfer

TIP   Technology, Innovation and Partnerships (Directorate)

UN   United Nations
UNOLS   University-National Oceanographic Laboratory System
USD   U.S. dollars

Summary

Every person on Earth is affected by the condition of the ocean. The ocean is shifting in unanticipated ways and at an unprecedented pace, with changes apparent at local, regional, and global scales. The ocean provides essential resources such as food, mode of transport, energy, minerals, medicines, and opportunities for recreation, all of which contribute to U.S. economic competitiveness and well-being for individuals and communities. It protects large parts of our national borders. The ocean also plays an important role in generating hazardous conditions that impact communities near coasts, where half the world's population resides, as well as communities farther inland.

Societal interests in the ocean—from economic activities and national security, food, recreation, and tourism—have garnered attention and support from citizens and legislators for decades. Funding for basic ocean science research through the National Science Foundation's (NSF's) Division of Ocean Sciences (OCE) since the 1950s has promoted U.S. world leadership in the field of ocean science and produced major scientific discoveries. For example, support from OCE has improved our ability to forecast El Niño events, providing up to 18 months' notice to prepare for and mitigate the effects of these and other ocean processes that dramatically impact the U.S., global weather, and economies. Support from OCE led to the discovery of hydrothermal vents as well as a massive microbial community deep below the seafloor, which have granted a window into the limits of life on Earth (and contributed to development of new biotechnologies). OCE-supported research increased our understanding of which marine species (including commercially relevant fish) are sensitive to changing ocean conditions, allowing for a more focused deployment of resources to sustain vulnerable species, including new efforts to conserve, restore, and future-proof coral reefs. Moreover, OCE's support of basic research on comb jellies led to discoveries of biomolecules that could lead to advances in neuroscience and treatments for diseases like Alzheimer's.

At the start of this new decade (2025–2035), U.S. investment in science, technology, engineering, and mathematics in general, and the geosciences in particular, is not keeping pace with growing societal needs. For instance, policymakers need better information on how ecosystem change influences important fisheries, how greater access to the Arctic will challenge U.S. national security, and how much more heat the ocean will absorb from the atmosphere as the climate continues to warm. At the same time, other countries, including U.S. competitors, are increasing their investments in ocean science and advancing their capacities, thus challenging U.S. scientific leadership. The United States is at a critical juncture, in which major investments are needed for upgrading and replacing infrastructure to support basic and applied research. Without such investments, the United States will lose preeminence in this globally competitive endeavor.

This report provides advice to NSF on focusing its investments in ocean research, infrastructure, and workforce to meet national and global challenges in the coming decade and beyond, and in doing so, enhance national security, scientific leadership, and a thriving blue economy.

THE COMMITTEE'S TASK

NSF asked the committee for advice on a research, infrastructure, and workforce development strategy that will advance understanding of the ocean's role in the Earth system from 2025–2035, and beyond. With the understanding that the Ocean Sciences Division at NSF (OCE) will continue to fund basic research across the ocean sciences, the committee was instructed to highlight a few research priorities for additional investment. The committee concluded that these priorities should include multi-sectoral and multi-disciplinary, use-inspired and societally-relevant components; they should provide opportunities to increase the workforce and broaden opportunities in ocean science, and these research priorities should enable new opportunities for OCE to work together with other units within NSF and with agencies and organizations outside of NSF, to address urgent questions surrounding ocean and Earth system science.

THE CHALLENGE FOR THE NEXT DECADE

Understanding and anticipating change in the ocean and how this change affects marine ecosystems and humans has never been more urgent. The committee poses the following challenge to NSF and to the broader ocean science community for the next decade: **by 2035, establish a new paradigm for forecasting ocean processes at scales relevant to human well-being.**[1] Establishing this paradigm will require a relentless focus on furthering basic science to observe and understand ocean processes through the lens of disciplinary, as well as transdisciplinary, innovative research practices. Progress on this research will contribute to federal management and policy decisions that promote adaptation to changes in the Earth system, resiliency, national security, and prosperity in the ever-changing environment and, it will position the United States as a leader in transformative ocean science.

OCEAN SCIENCE RESEARCH PRIORITIES

To truly increase fundamental understanding of ocean processes to the point of being able to predict them in areas that are of vital importance to society, the committee recommends that investment be focused on facilitating and promoting research around three main themes and associated urgent research questions: Ocean and Climate, Ecosystem Resilience, and Extreme Events, described in the following sections.

Ocean and Climate

The ocean is presently absorbing 90 percent of the heat and approximately 30 percent of the carbon that result from global emissions of greenhouse gases. Any decline in these rates of uptake would accelerate carbon dioxide (CO_2) accumulation and warming in the atmosphere. In essence, the ocean has served as a buffer, absorbing much of the excess CO_2 emitted since the industrial revolution and thus lessening the degree of change seen in Earth's climate. It is unlikely that the ocean will continue this same rate of absorption, and it is possible that the patterns of ocean circulation that regulate Earth's climate will shift. Large changes in ocean circulation will influence hurricane and storm development and movement; ice sheet stability; ocean chemistry, including acidification, nutrient speciation and cycling; and ocean productivity and ecosystem health, including major die-off events (e.g., coral bleaching). These changes could be gradual, or more sudden and dramatic, or even irreversible. Additional research will be required to forecast such changes and inform society on the anticipated conditions in the coming decades. Thus, the first urgent research question is *How will the ocean's ability to absorb heat and carbon change?*

Research targeting this question is wide-ranging and could include topics such as developing new approaches to observe heat transport, improving predictions of marine ice sheet instability, and developing ways to quantify variability in the carbon cycle. Potential outcomes of this work include improved forecasts

---

[1] Forecasting the state of the ocean at scales relevant to human well-being must occur at various temporal and spatial scales as dictated by the societal questions at hand and by the nature (e.g., rates of change, response times) of the ocean system processes.

of potential tipping points in the Earth system, the ability to predict ocean uptake of CO_2 and thus assess the potential for climate mitigation strategies such as marine CO_2 removal.

Ecosystem Resilience

Fundamental changes in both the Earth and ocean's systems are resulting in shifts to ecosystems that may negatively impact the local and global communities that depend on them. Humans could lose not only direct benefits they derive from the ocean (e.g., food and other biological products, coastal protection), but also benefits of marine ecosystems in the global earth system (e.g., carbon and nutrient cycling), which affect human health and well-being. Forecasting the resilience of marine ecosystems to these ecosystem shifts remains a significant challenge. Reducing the scientific uncertainty around ecosystem change would enhance communities' ability to adapt and maintain functionality. Therefore, the second urgent research question is *How will marine ecosystems respond to changes in the Earth system?*

Examples of research directions that support this question include determining the effects of warming, acidification, and de-oxygenation on the dynamics and productivity of ocean biota, and co-developing tools for rapid measurements and assessments of ocean biological and functional diversity. Development of food web models that can predict food security, assessment of potential ecosystem effects of deep-sea mining, and design of effective area-based conservation measures are just a few of the societally relevant outcomes of such research.

Extreme Events

While the ocean provides many benefits to society, it is also the source of devastating events such as earthquakes, tsunamis, hurricanes, storm surges, and flooding that directly impact coastal communities. In addition to these coastal hazards, increased precipitation, atmospheric rivers, and heatwaves have far-reaching impacts on coastal and inland communities. Long-term hazards, such as saltwater intrusion and land loss from sea level rise, currently threaten coastal communities and are expected to become increasingly more severe. Societal vulnerability to these extreme events can be profound, including its dependence on massive, century-scale investments such as port facilities, existing commercial fishing fleets, offshore energy infrastructure, coastal development, and national defense infrastructure. In light of this dependence on the ocean, and the socio-economic and life-saving benefits of forecasting extreme events, improving the ability to observe, understand, and thereby forecast extreme events across spatial and temporal scales relevant to human well-being is critical. To facilitate societal adaptation in the United States and across the globe, the third urgent ocean science question is *How can the ability to forecast extreme events driven by ocean and seafloor processes be improved?*

Example research directions in support of this theme include integration of knowledge to improve early warning systems of geohazards, increasing ability to predict global weather extremes and apply these forecasts to inform sustainable urban planning, agricultural and forestry practices, and coastal community climate mitigation strategies. Ultimately, the research required to improve forecast of extreme events will improve the safety of communities vulnerable to extreme weather and geophysical hazards.

> **RECOMMENDATION 2.2: The National Science Foundation's Division of Ocean Sciences should support basic research that addresses the goal of forecasting ocean processes at scales relevant to human well-being, with emphasis on the following themes and questions:**
> - **Ocean and Climate: How will the ocean's ability to absorb heat and carbon change?**
> - **Ecosystem Resilience: How will marine ecosystems respond to changes in the Earth system?**
> - **Extreme Events: How can the ability to forecast extreme events driven by ocean and seafloor processes be improved?**

This compelling, cross-cutting research portfolio can galvanize the ocean science community to work together toward the challenge while offering opportunities to build new partnerships and increase interest in multiple aspects of oceanographic research.

WHAT WILL IT TAKE?

NSF is the only U.S. federal funding agency founded with the explicit mission to sponsor basic research. For the field of ocean sciences, basic research includes improving fundamental understanding of the ocean and its interactions with other parts of the Earth system. Basic research underlies progress across the range of applied and solutions-oriented work in the ocean sciences.

> **RECOMMENDATION 2.1: The National Science Foundation's Division of Ocean Sciences should continue to support a broad portfolio of basic research to ensure that scientists and engineers of the future have access to a continually improved understanding of the ocean to advance and fuel innovation and resilience in the United States.**

Accelerating U.S. scientific progress in understanding and forecasting ocean systems and processes will require an integrated approach that takes full advantage of emerging technologies and methods along with new resources. This includes embracing new artificial intelligence and machine learning techniques; reinvesting in ocean science infrastructure and core science research; curating vast sets of ocean data; developing novel observational technologies that fully utilize advances in robotics, environmental genomic analysis, and acoustics capabilities; and thinking creatively about how to utilize existing underwater infrastructure, such as communication cables, for research. Accelerating ocean science knowledge will also require international collaborations, U.S. public–private partnerships, thoughtful consideration of open and accessible data management, and broadening participation to ensure that all ways of knowing are incorporated. In short, the field of ocean science must transform so that research advancements are accelerated and can be applied rapidly to the urgent issues that the United States and our planet face now and in the near future. OCE continues to play a crucial role in funding these efforts (Figure S.1).

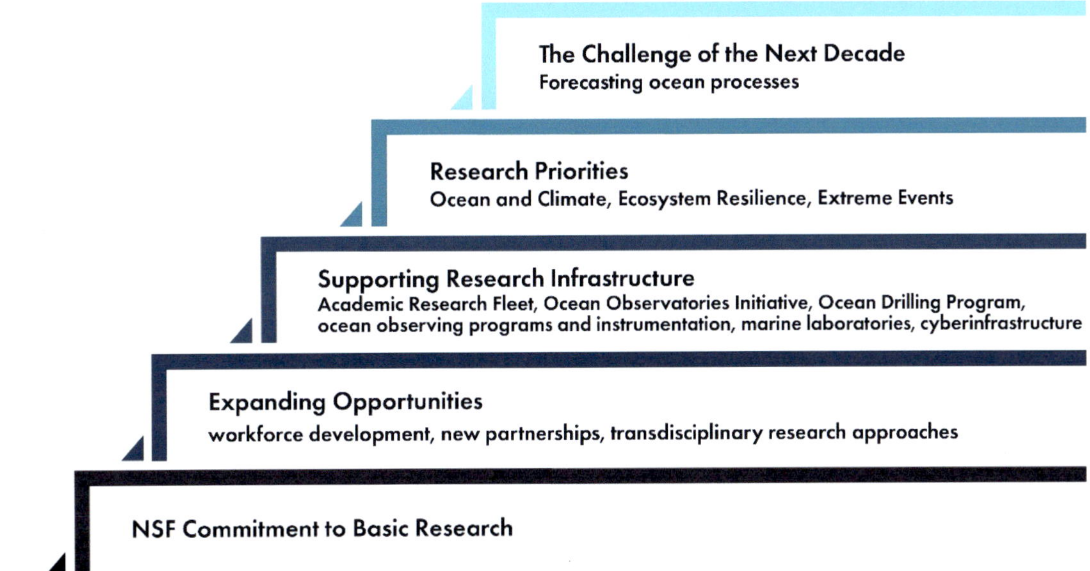

FIGURE S.1 Path from basic research to meeting the challenge of the next decade.

EXPANDING OPPORTUNITIES

Supporting and integrating the research needed to achieve the research priorities of the next decade require a transformation within ocean science into a field that is more collaborative, leveraging both intellectual and financial resources. To accomplish this, the committee recommends that OCE commit to expanding opportunities for science in a manner that promotes transdisciplinary research, expands the workforce, and includes new partnerships.

Transdisciplinary mindsets, skillsets, and practices—including collaboration with scientists in other disciplines, such as social sciences, and with local knowledge holders—will be key to actualizing the new paradigm of ocean science. Including multiple perspectives, among not only ocean science researchers, but also ocean science funding agencies and their directorates, will advance innovation and application of basic research. The path toward conducting successful transdisciplinary ocean science research has been partially exemplified, as evidenced by research projects in ocean acidification, marine heatwaves, and hypoxia, but it is not easily put into practice. For example, understanding ocean acidification, a chemistry problem that is deeply intertwined with the ecosystem and society, requires a shift in standard research practices to bring together differing knowledge holders to conduct research on complex interactions of multiple stressors.

In recent years, NSF has recognized the importance of supporting transdisciplinary science, as evidenced through their support of new projects such as the Cascadia Coastlines and Peoples Hazards Hub. Going forward, resource investment is needed, both fiscally and programmatically (including leadership development, management, and outcome assessments), to continue to support and incentivize building and sustaining transdisciplinary research teams in the field of ocean sciences.

RECOMMENDATION 3.1: To foster transdisciplinary research that promotes emerging solutions to challenges related to changes in ocean systems and processes, the National Science Foundation's (NSF's) Division of Ocean Sciences should invest in projects that utilize a participatory process toward relationship-building and collaborative efforts, establishing long-term trust and knowledge-sharing and impacting broad interests. NSF should explicitly enable research that crosses directorates and programs and intentionally facilitate efforts to dismantle barriers to transdisciplinary approaches and implementation. Potential strategies include:

- **supporting projects with measurable social and environmental returns on investments, especially use-inspired and solutions-oriented research on the changing ocean;**
- **investing in training an expanded ocean science workforce that includes developing essential science, technology, engineering, and mathematics skills as well as transdisciplinary skills that enable meaningful connections to the humanities, social sciences, and economics;**
- **implementing requirements for proposals that encourage partnerships with interest holders including local communities, regional organizations, Indigenous and Tribal groups, and others, to participate in ocean research; and**
- **financially supporting platforms and networks that facilitate knowledge exchange between interest holders, sectors, and disciplines.**

Successfully addressing ocean science research challenges in the coming decade will require the cultivation and development of an expanded ocean science workforce. The workforce will benefit not only from high-quality scientific training, but also from training in (1) collaborative science methodologies to engage with and learn from those who have differing perspectives and knowledge systems, (2) best practices for engaging in use-inspired research, (3) pathways to interacting with industry partners and interested communities, and (4) collaboratively developing and implementing research that addresses societal needs. Investing in training for transdisciplinary skills can result in a workforce prepared to succeed not only in the academic sector of ocean sciences but also more broadly in solutions-oriented science and other sectors of the U.S. economy. For example, opportunities to participate in an oceanographic research cruise, or work remotely with robotic systems, can catalyze an early-career individual to build a varied set of skills with wide application.

RECOMMENDATION 3.2: The National Science Foundation's Division of Ocean Sciences (OCE) is uniquely positioned to shape the future of ocean sciences research and policy by cultivating a workforce that includes multiple skillsets and knowledge systems. OCE should:
- **Explicitly support reskilling and upskilling, as well as mentorship, of the academic ocean science workforce to better engage with industry, entrepreneurs, interest holders, and other partners, to promote leadership development in seagoing technologies, data management, cultural competencies, and educational and mentoring practices.**
- **Support workforce development by investing in vocational and academic pathways, such as offering scholarships and apprenticeships; establishing fellowships with thoughtful considerations of metrics that support underserved student and professional populations, including students, researchers, or interest holders in the locations where the research is being conducted; and funding and incentivizing collaborative projects that bring together researchers from varied geographic and disciplinary backgrounds.**
- **Promote safe working environments by enforcing and incentivizing policies that protect people from discrimination, harassment, and bullying.**

Partnerships with other sectors can reduce the expense of working in the ocean. New, creative partnerships could engage multiple sectors and foster a broader, more collaborative approach to ocean and Earth science. Development of strategic partnerships will be both necessary and critical to accomplishing the ambitious and urgent research portfolio put forward, specifically in the area of ocean observing (including platforms, data, and technology development) across ocean physics, chemistry, biology, and geology. Partnerships will also promote opportunities to engage with knowledge systems and communities across different disciplines and sectors with interests in the health of the ocean.

RECOMMENDATION 3.3: To bolster and leverage urgent ocean science research over the next decade, National Science Foundation's (NSF's) Division of Ocean Sciences (OCE) should explore various ways of bringing new resources to this work, including expanding partnership

with other NSF directorates, related mission agencies, industry, and other organizations. OCE should seek greater interagency cooperation through federal policies and mechanisms for leveraging NSF-funded basic ocean science research with those mission agencies. Developing new partnerships in ocean sciences, as well as other disciplines, may increase support available for the development of new technologies, data curation, solutions-oriented research, and the basic research that will fuel these developments.

Supporting Research Infrastructure

Regaining U.S. leadership in ocean science will depend on physical infrastructure for observing and studying the ocean to answer the questions of the next decade. The committee recommends continued funding for core OCE-supported infrastructure in support of the three research priorities, as well as investment in new and emerging observational technologies and cyberinfrastructure. The importance of investment in new and improved technologies for ocean observing and computing cannot be overstressed.

Academic Research Fleet

The Academic Research Fleet (ARF), including the National Deep Submergence Facility (NDSF), remains a vital part of collecting oceanographic data at sea and is required infrastructure for basic ocean science research and for research related to the urgent ocean science priorities described in this report. The ARF also plays an important role in ocean science workforce development by providing support for early-career training onboard vessels.

In the coming decade, research vessel capacity will be significantly reduced, owing to the end-of-life of multiple research vessels across the intermediate, global, and local classes. The overall estimated losses in ship time will be partly mitigated when the new NSF-owned regional-class research vessels come online. However, they are not replacements for the global-class vessels that can operate for longer periods at sea, operate in more challenging sea states, and carry a heavier load of instrumentation than smaller vessels. As several ships near their end-of-life, replacements with comparable or better capabilities are necessary to accomplish high priority science. Without replacing the global-class vessels, U.S. leadership in ocean science will be vastly diminished.

RECOMMENDATION 4.1: The National Science Foundation's (NSF's) Division of Ocean Sciences and appropriate partners should conduct an evaluation that details the funding allocation for the Academic Research Fleet in order to make informed distributions of the limited resources available over the next decade and beyond. The evaluation should consider metrics such as days of use funded by NSF; use of local vessels for NSF-funded research; geographic distribution of assets; and alternative mechanisms for providing researchers with access to the coasts (including salt marshes) and sea, such as through marine laboratories or new partnerships.

The long-term status of global-class research vessels is concerning; the committee urges interested parties to initiate and/or continue significant and intentional planning for replacements. Without such planning and subsequent implementation, accomplishing high-priority basic and solutions-oriented science will be compromised, and U.S. leadership in providing access to the ocean will be curtailed.

Ocean Observatories Initiative

NSF's Ocean Observatories Initiative (OOI) has seen many successes over the past decade and as it stands, would partially support the urgent research questions for the next decade such as forecasting extreme events and providing data critical to forecasting changes in circulation, carbon and heat transfer, and understanding changes to fisheries and marine ecosystems. However, OOI was not designed to align with specific science needs and would benefit from a revisioning and restructuring exercise that takes into consideration the current and future needs of the ocean science community broadly and how best to meet those needs.

RECOMMENDATION 4.2: The National Science Foundation's Division of Ocean Sciences should conduct a revisioning and restructuring exercise for the future of the Ocean Observatories Initiative (OOI). The review, which should occur as a fully separate activity from the usual renewal/recompete discussions, could include:
- **an analysis of the scientific contributions of each of the OOI arrays;**
- **reconsideration of the goals and objectives of the program to better address the needs of the ocean science community and align with the evolving and urgent ocean science questions for the next decade; and**
- **consideration of how to incorporate technology that may not have existed when OOI was originally envisioned, including innovative ways to observe and measure biological abundances and processes, such as low-cost distributed observational networks.**

This recommended review differs from, and should be independent of, the renewal process that is focused on how well the infrastructure is meeting the requirements of the existing cooperative agreement. Rather, this investigative approach would provide a timely path forward for the next iteration of OOI that takes advantage of strategic partnerships, the evolution of ocean sciences and technological capabilities over the two decades since the OOI was originally conceived, and the approaches needed to address the urgent research priorities laid out in this report, as well as other science questions. This recommended review should occur well before the initiation of the review/renewal discussion regarding the current cooperative agreement.

Scientific Ocean Drilling

For nearly 60 years, scientific ocean drilling has played a pivotal role in our understanding of the Earth system, such as elucidating theories on plate tectonics and ice sheet behavior and constraining climate models. Going forward, scientific ocean drilling remains an important tool for answering the urgent science questions of the next decade around the topics of climate forecasting, carbon and heat cycling, subseafloor microbial communities, and forecasting geohazards. However, the contract for the primary drilling platform supported by the United States since 1985, the *JOIDES Resolution,* expired in 2024. As such, the future of the scientific drilling program is unknown. Efforts to develop plans for a new vessel—and in the meantime, to secure mission-specific drilling vessels to continue the work of this legacy program—are seen as key steps forward. Without new approaches to addressing both legacy assets and infrastructure needs, the capacity for future U.S. operational and scientific leadership is bleak. It is unlikely that such U.S. leadership can be regained without a U.S.-based drillship. Dedicated funding will be needed to maximize research enabled by the use of legacy assets and to plan for a new era of scientific ocean drilling which includes U.S. leadership and embraces international collaboration.

RECOMMENDATION 4.3: The National Science Foundation (NSF) should take action to regain U.S. leadership in scientific ocean drilling on a global stage. To support basic ocean science and the urgent ocean science research portfolio identified in this report for the next decade and beyond, the committee makes the following recommendations:
- **Legacy assets: NSF's Division of Ocean Sciences (OCE) should create a dedicated funding line for the Legacy Asset Projects program to support expedition-scale collaborations, which maximize the return on legacy assets by providing funding to scientists to conduct large-scale research with existing cores and data.**
- **New drilling infrastructure:**

- o OCE should develop a sustainable ocean drilling program, taking into account the need for a U.S.-based drillship. The committee urges NSF to creatively implement management and operational models that ensure viable long-term operations of such a vessel, including developing and nurturing more options for industry work. Such new management and operational models may differ significantly from the recently concluded drilling program.
  - o Until a dedicated drillship is available, OCE should continue to support the use of mission-specific platforms for addressing high-priority, urgent science questions, and to the extent possible, support efforts to retain the technical and engineering expertise in deep-sea coring previously employed by the United States with the *JOIDES Resolution.*
- **International collaboration and coordination:**
  - o NSF and international funding agencies and governments should coordinate and collaborate globally toward an integrated, long-term strategy for scientific ocean drilling. Such collaboration will require meaningful and transparent reciprocity in scientific participation levels, financial support, scientific planning, and more among contributing partners.
  - o OCE should renew dialogue with both new and long-standing international partners to identify cost-sharing models for dedicated drillship operations; such models may include scientist shipboard participation proportional to international contribution levels.

Additional Supporting Infrastructure

In addition to supporting the ARF, the OOI, and ocean drilling, OCE supports several other programs that provide the infrastructure to collect data that is necessary to answer the urgent ocean science research questions of the next decade such as the Ocean Bottom Seismographic Instrument Pool, Marine Rock and Sediment Sampling, Ocean Data Facility, Mooring Facilities and Services, autonomous platforms, and marine laboratories. Existing infrastructure is not enough; continued development and/or adaptation of new tools and technologies will enable the sampling of the depths of the ocean and seafloor. Technological development should be supported and conducted in collaboration with strategic partners, involving use-inspired focus groups, community engagement, and broad access for participation and input by transdisciplinary groups.

RECOMMENDATION 4.4: In addition to continuing funding for existing facilities as they support ocean science research priorities for the next decade, the National Science Foundation's Division of Ocean Sciences, in cooperation with agency and other partners (including private philanthropies), should explicitly support:
- **the increased use of autonomous assets for making sustained observations and increasing the efficiency of oceanographic expeditions (expanding the footprint of the vessel while at sea);**
- **new mechanisms for providing researchers with access to instrumentation—for example, shared instrument pools;**
- **the collaborative development of revolutionary and innovative sensor technologies for such efforts as collecting ocean chemistry data (e.g., partial pressure of carbon dioxide), biological data (e.g., environmental and organismal DNA, bio-sensors for species abundance), and measuring seafloor geodesy;**
- **new avenues for bringing novel sensors to market at scale and broadening access to the larger research and management community;**
- **expansion of data curation efforts to support bioinformatics, artificial intelligence, other analyses, and modeling; and**

- **the evaluation of new applications of acoustics (e.g., distributed acoustic sensing), Science Monitoring and Reliable Telecommunications cables, and other emerging technologies to identify their potential to contribute to future research.**

Universal to all oceanographic assets, data curation that follows FAIR (findable, accessible, interoperable, and reusable) and CARE (collective benefit, authority to control, responsibility, ethics) principles is essential for the collaborative development and access of more accurate ocean and seafloor forecasting models, especially when utilizing advances in machine learning and artificial intelligence. The collection and effective integration of disparate ocean datasets—especially new datasets from emerging technologies, such as biotechnology, to support artificial intelligence and machine learning—are needed to create a unified and interoperable ocean information system. Much can be gleaned from existing and recent national and international efforts to develop robust ocean data management strategies; a workshop series or other convening activity would be an appropriate next step for OCE.

RECOMMENDATION 4.5: The National Science Foundation's Division of Ocean Sciences should fund a convening activity, such as a series of workshops, that seeks to gather expert advice and input, review established strategies, and develop peer-reviewed guidelines and practices for ocean science data curation, computing, and security, both on research vessels and on shore, integrating findable, accessible, interoperable, and reusable (FAIR) and collective benefit, authority to control, responsibility, ethics (CARE) principles and the required supporting cyberinfrastructure. Data and computational experts from adjacent fields (e.g., applied mathematics) should be encouraged to participate.

CONCLUSION

The 2025–2035 Decadal Survey was developed to ensure that OCE will be empowered to continue to provide the foundational support for ocean science research, not only within NSF, but also across the federal agencies that use ocean science research to accomplish their missions. Now is the time for the United States to invest and take leadership in answering the urgent ocean science priorities outlined in this report, allowing researchers, policy makers, and other leaders to anticipate and understand changes in the ocean system. Such understanding is critical for national security, economic prosperity, environmental stewardship, and for the well-being of humans and the ecosystems on which they depend, in the next decade and beyond.

1

Introduction

This report was developed to advise the National Science Foundation's (NSF's) Division of Ocean Sciences (OCE) on a research, infrastructure, and workforce development strategy for advancing the understanding of the ocean's role in the Earth system from 2025 to 2035 and beyond. Although the recommendations are specific to OCE, many of the actions will be made more impactful by the involvement of other agencies and organizations as research advances are used to tackle ocean-related problems and opportunities.

The ocean is a life-sustaining reservoir that connects humankind and provides essential resources such as food, energy, minerals, medicines, and recreation opportunities, all of which contribute to U.S. economic competitiveness and well-being for individuals and communities. The ocean also generates hazardous conditions that impact communities near coasts, where half of the world's population resides, as well as those farther inland, which are impacted indirectly. Thus, every person on Earth is affected—at times positively and other times negatively—by the condition of the ocean. This is emphasized in the following headlines from 2024 that refer to events directly influenced by the ocean:

"Cold Weather Businesses Suffer in the Winter That Wasn't"—*The Wall Street Journal*, March 8 (Carlton, 2024)

"'Rivers in the Sky' Have Drenched California, and yet Ever More Extreme Rains are Possible"— *Los Angeles Times*, April 25 (Toohey, 2024)

"The Drowning South: Where Seas are Rising at Alarming Speed"—*The Washington Post*, April 29 (Mooney et al., 2024)

"Ocean Temperatures Surge, Threatening Worst Coral Bleaching Event in History, Scientists Say"— *Fox News*, May 17 ("Ocean Temperatures…", 2024)

"Storms Again Strike Texas, Leaving Hundreds of Thousands Without Power"—*The Wall Street Journal*, May 29 (Pisani, 2024)

"Unprecedented Ocean Temperatures Make This Hurricane Season Especially Dangerous"—*USA Today*, June 2 (Pulver, 2024)

"Watch the Glacier Outburst That Sent a Surge of Water Into Juneau, Causing 'Unprecedented' Flooding"—*CNN*, August 7 (Zerkel, 2024)

"How Close are the Planet's Climate Tipping Points? Earth's Warming Could Trigger Sweeping Changes in the Natural World That Would Be Hard, if not Impossible, to Reverse"—*The New York Times*, August 11 (Zhong and Rojanasakul, 2024)

"A Climate-Related Mass Die-Off Leaves Over 100 Tons of Dead Fish Collecting at a Greek Port"— *Associated Press News*, August 30 (Kousioras and Gatopoulos, 2024)

"The Hottest Summer on Record Could Lead to our Warmest Year Ever Measured. Again."—*PBS*, September 6 (Borenstein, 2024)

These headlines all describe events that were influenced by the ocean; they point to the importance of the ocean to human well-being and demonstrate the need to better understand how the ocean functions. Coastal communities and infrastructure are impacted by increased flooding events, both extreme and chronic. The repercussions of these impacts range from impaired human health and direct injury to economic losses. Warming ocean waters are also driving extreme weather across the globe, impacting lives, economies, and national security. Rising sea levels (Wong et al., 2017), altered circulation patterns, changes in the acidity of ocean water, and geographic shifts in ocean ecosystems (Garcia-Soto et al., 2021) are all being observed. In addition, the ocean feeds and provides key nutrients to hundreds of millions of people; but the sustainability of this food source is uncertain, as seafood and shellfish industries are impacted by the acidification of the ocean waters, the presence of harmful algal blooms, and the emergence of low-oxygen zones. Adaptation and mitigation strategies are needed to preserve or even enhance the livelihood of communities being affected.

Both basic and applied research in the field of ocean sciences are key to developing more accurate forecasting of ocean and seafloor processes that affect the lives and well-being of humans and all other species on the planet. This forecasting can enable society to realize emerging ocean-oriented opportunities for food sources and energy security. Scientific progress over the last few decades has been significant, fueled in part by advances in observational technologies and increased availability of computational resources. Yet the need is urgent to further accelerate scientific progress and rapidly deploy the new understanding to advance forecasting capabilities. NSF's funding of basic research, coupled with enhanced partnerships with mission agencies, can make this vision a reality.

Producing and validating forecasts is an excellent way to test the level of understanding of an ocean process and its effects on the Earth system. For example, forecasts of changes to surface salinity over large areas of the ocean can be used to determine rates of evaporation and ultimately to make long-term (monthly) forecasts of continental rainfall patterns (Li et al., 2016). In fact, these forecasts are so accurate that Salient,[1] a for-profit company, can market its long-term rainfall forecasts for agricultural and other applications. This example is not yet typical of ocean sciences, and accurate forecasts cannot yet be produced for many societally important phenomena, owing to a lack of understanding of causal linkages. For example, following a series of heat waves between 2018 to 2021, the snow crab fishery in the eastern Bering Sea collapsed unexpectedly (Szuwalski et al., 2023). The ability to forecast such an event would have aided in mitigating the impact on the fishing economy. Another example involves Atlantic meridional overturning circulation (AMOC), the dominant circulation system in the North Atlantic, which includes the Gulf Stream. AMOC affects climate and sea level on both sides of the Atlantic. Questions as to whether this important circulation system is slowing remain unresolved (e.g., Rahmstorf, 2024; Roquet and Wunsch, 2022). Much depends on the answer, as an AMOC slowdown has potentially serious consequences for the United States and other countries bordering the North Atlantic. For a final example, results from highly sensitive measurements are taken from sensors placed in boreholes by ocean drilling vessels, active seismic measurements from oceanographic vessels, and passive seismic and geodetic measurements from ocean bottom seismometers and other instruments deployed on the seafloor; these measurements are revealing the precursors to significant ocean earthquakes. Sustaining and improving these measurements raises the possibility of more accurate forecasts for the location and timing of seafloor earthquakes.

Moving from understanding ocean processes to forecasting them is an important step that could have significant societal benefit. U.S. scientific progress in these areas will require embracing new tools from computational science, artificial intelligence (AI), and machine learning; reinvesting in ocean science infrastructure and core science research; curating vast sets of ocean data; developing new sensors; fully utilizing advances in robotics, environmental genomic analysis, and acoustics capabilities; and thinking creatively about how to utilize existing underwater infrastructure, such as communication cables, for science. Accelerating ocean science knowledge also requires international collaborations, U.S. public–private partnerships, thoughtful consideration of data management, and broadening participation to ensure that all ways

---

[1] See https://salientpredictions.com.

of knowing are incorporated. In short, the field of ocean science must transform so that research advancements are accelerated and can be applied rapidly to the issues that the United States and our planet face now and in the near future. OCE continues to play a crucial role in these efforts.

This report highlights forward-looking areas of basic ocean science research that will advance the understanding of the ocean in ways that can lead to more accurate forecasting of ocean conditions and their impact on all forms of life. Note that the basic research that would lead to more accurate forecasting would clearly fall under OCE's purview, while incorporating the resulting new understanding into operational forecasting models would be carried out by other federal agencies or through private industry. The report also identifies new and innovative ways of conducting basic research to improve the ability to forecast the future ocean, and in doing so, create the workforce and partnerships needed to succeed—this is an "all hands on deck" approach.

THE IMPORTANCE OF OCEAN SCIENCE TO THE NATION AND THE ROLE OF NSF

The ocean's importance to national security, leadership in science, a prosperous economy, and the health and well-being of humans has long been recognized by U.S. taxpayers, garnering bipartisan funding and policy support for decades.

National Security and Scientific Leadership

Federal funding of ocean research and technology began in the mid-1800s with the establishment of the Naval Observatory and Hydrographic Office; this funding was accelerated during World War II through the U.S. Navy. Knowledge of the ocean is critical to defending the country's coastlines and those of allies, and knowledge of the weather affects all defense operations.

In this century, the changing ocean is already affecting national security. For example, the Department of Defense (DOD) noted that the dramatically altered natural environment of the Arctic is "creating a new frontier of geostrategic competition" (DOD, 2021, p. 6), and that more than 1,700 DOD sites could be affected by sea level rise (DOD, n.d.). DOD also recognizes that it "relies on authoritative scientific data and modeling provided by USG [U.S. Government] science agencies," such as NSF (DOD, 2021, p. 14). Since its establishment in 1950, NSF has supported academic research on a broad set of ocean topics.

In 2011, members of the U.S. Senate formed an Oceans Caucus to stress the importance of the ocean and its crucial role in all aspects of life. Additionally, the White House established the Subcommittee on Ocean Science and Technology (SOST) to coordinate federal actions related to ocean issues; this group issued the report *Science and Technology for America's Oceans: A Decadal Vision* (SOST, 2018), which identified five goals for the decade of the 2000s, including (1) understanding the ocean in the Earth system and (2) developing resilient coastal communities. The goals were emphasized in the 2023 *Ocean Climate Action Plan* (OPC, 2023a), among other efforts.

For many decades, NSF's OCE has supported ocean science research—primarily basic research at academic institutions. This research has provided fundamental new insights into the ocean's role in the Earth system. OCE also provides most of the operating funds for the Academic Research Fleet (ARF); operates the mooring arrays and other equipment of the Ocean Observatories Initiative; and, through 2024, operated the *JOIDES Resolution,* the only deep-ocean drilling vessel widely available to the U.S. research community and its international partners. OCE-sponsored research spans the global ocean, enabling the U.S. oceanographic research community to lead the world in a breadth of ocean discoveries.

Today, the U.S. ocean science enterprise is at a crossroads, and U.S. leadership in this field is in jeopardy. In 2024, NSF announced the cessation of the cooperative agreement to manage the *JOIDES Resolution,* leaving the United States without the capability to collect cores in most of the global ocean and challenging the country's multidecadal leadership in this field. The rest of the ARF is also aging, and the new fleet of regional-class research vessels will not begin operations for several years. Meanwhile, other countries, such as China, are investing heavily in ocean science infrastructure, particularly as it relates to deep-sea exploration (Normile, 2024; Varbanov, 2023). In combination, these events leave the United

States without the infrastructure needed to support its leadership in science, technology, and engineering (NASEM, 2024a, 2024b, 2024c; NSB, 2024).

The Blue Economy

The ocean supports important resources and services—including food, energy, mineral resources, medicines, transportation and trade, and recreation—that benefit humans and society and overall U.S. economic competitiveness. The *blue economy* is the sum of the economic activities of ocean-based industries, together with the assets, goods, and ecosystem services provided by marine environments.

In 2018, the ocean supported roughly 2.3 million jobs in the United States, accounting for 1.5 percent of total U.S. employment (Bureau of Economic Analysis, 2020). And, as of 2022, the annual blue economy was estimated at 476.2 billion U.S. dollars (USD) (Bureau of Economic Analysis, 2024). Globally, the blue economy is projected to reach at least 3 trillion USD and to double its growth rate by 2030 (OECD, 2016). This sets the global blue economy as one of the seven largest economies in the world. A sustainable and thriving blue economy will require strategic management practices informed by improved understanding and forecasting of ocean processes.

Health and Well-Being

The ocean strongly influences conditions, including the climate, for human life on Earth. Since the industrial revolution, the ocean has absorbed 90 percent of excess heat, serving a major role in modulating the climate and making more of Earth's land habitable—although this buffering service has come at a cost. Global, regional, and local ocean conditions impact communities in significant ways, and as the ocean becomes warmer and fresher (with increased glacier and sea ice melt), the impacts are becoming more severe and are compounding.

Warming waters can lead to increasing intensity of storms forming over the ocean, which bring high wind and wave action and heavy rainfall (including far inland, as seen in the mountains of North Carolina during Hurricane Helene in 2024) that can have catastrophic impacts on people, infrastructure, and habitats. The Pacific Islands—including but not limited to the Hawaiian Islands, Samoa, Guam, the Marshall Islands, and Micronesia—are among those areas that will be vulnerable earliest, as are communities in Alaska, which also experience the impacts of melting permafrost and loss in ice cover that can otherwise protect the coastlines from the destructive impacts of storms. Warming waters are also causing rising sea levels, which further amplify the impacts from the storms as well as intrusion of saltwater into groundwater and impacting freshwater supply in aquifers.

Globally, the ocean's coastline is home to roughly half the human population. Nationally, U.S. coastal zones (<100km to ocean) represent less than 10 percent of the U.S. land area yet are now home to about 40 percent of residents, with a growth rate from 1970 to 2020 of 46 percent (NSF, 2024d). These communities are directly connected with the ocean and are acutely impacted by increased storms, coastal flooding, and atmospheric rivers dropping increased amounts of rainfall (Rhoades et al., 2020). Their livelihoods and, in many cases, their very existence depend on responsible stewardship of the ocean and its resources, informed through different knowledge systems, ocean science research, and development of decision-making tools. For example, fishing communities and the commercial fishing industry are being impacted by hazards related to shifting fish stock distributions, degradation of suitable habitats, ocean acidification, harmful algal blooms, and low-oxygen zones (Gobler and Baumann, 2016).

The crucial role the ocean plays in global health and human well-being is embodied in the 2021 declaration of the United Nations Decade of Ocean Sciences for Sustainable Development (2021–2030), "by providing natural and innovative solutions to global challenges, from climate change to poverty eradication, the development of ocean sciences is essential for social, economic and environmental balance of the planet".[2] Society needs accurate forecasts of many important and often dramatic changes occurring to the

---

[2] See https://www.unesco.org/en/underwater-heritage/un-decade.

Earth system, including ocean and seafloor processes, that affect the life and well-being of all species on the planet. Basic research to support such work is critical.

STUDY CONTEXT

Decadal surveys conducted by the National Academies of Sciences, Engineering, and Medicine offer an opportunity for the science community to gather and reach consensus about science priorities for the next decade in a given field. In many cases, they include assessments of progress made towards research priorities identified in the previous decadal survey for that field. The present survey, referred to as the 2025 Decadal Survey, addresses ocean science research priorities for 2025–2035. It builds on and assesses progress since the 2015–2025 survey, referred to as the 2015 Decadal Survey, and its report, *Sea Change* (NRC, 2015), and also incorporates conclusions from the 2025 Decadal Survey interim report, released in 2024, focused on research and infrastructure priorities for scientific ocean drilling (NASEM, 2024b; Figure 1.1). See Box 1.1 for definitions of important terms used in this report.

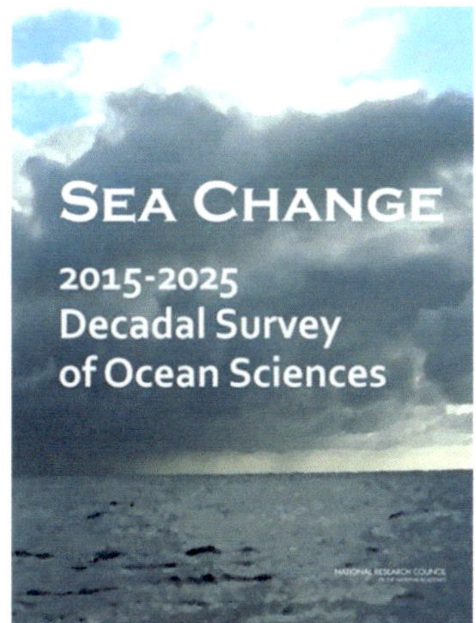

FIGURE 1.1 Publications resulting from decadal surveys of ocean sciences for the National Science Foundation. NOTES: Reports include the 2015 *Sea Change* report, the 2024 interim report *Progress and Priorities in Ocean Drilling*, and the 2025 report *Forecasting the Ocean*.

2015–2025 Decadal Survey of Ocean Sciences

The ocean science community engaged with the concept of a decadal survey for the first time with the release of the report *Sea Change: 2015–2025 Decadal Survey of Ocean Sciences* (NRC, 2015; see Figure 1.1), produced at NSF's request. At the time, there was intense concern about the portion of OCE's budget that was being used to support research infrastructure compared with funds spent on individual or collaborative research projects and programs (the "core science" programs). More specifically, in 2012, approximately half of OCE's budget supported research infrastructure, with the trend at the time indicating that operation and maintenance costs of infrastructure would continue to increase. The absence of significant increases in the OCE budget would result in a net reduction of funding for future research programs.

The 2015 Decadal Survey committee ultimately recommended that, during times of flat or reduced budget, "infrastructure expenses should not be allowed to escalate at the expense of core research programs" and use of a fixed ratio for infrastructure costs relative to the total budget would help ensure that one part of the budget does not increase at the expense of the other (NRC, 2015, p. 4). After receiving the 2015 Decadal Survey report, NSF balanced the budget between research and infrastructure through economized and prioritized infrastructure spending. This involved decisions to cut infrastructure costs from three main areas: (1) the Ocean Drilling Program: funding for the U.S. Science Support Program was decreased, with a request for more support from international partners; (2) the Ocean Observatories Initiative: two Southern Ocean moorings were removed from the portfolio as part of strategic reductions in certain operations; and (3) the ARF: funding was limited for the operation of the *R/V Marcus G. Langseth*. The result of these budget cuts was the transfer of funds from infrastructure to OCE core research, creating the even split between infrastructure costs and funding of research, as recommended. That balance has remained, more or less; however, the last few years have seen increased infrastructure costs, particularly due to operating an aging ARF, while the total OCE budget has remained the same.

BOX 1.1
Important Terminology Defined

Basic research provides the basis for solutions-oriented or applied science, even if the application of the research is not immediately known.

Vital scientific research is compelling, high-priority research with the potential to transform scientific knowledge of the interconnected Earth system and the critical role of the ocean in that system. Vital scientific research can lead to paradigm shifts in understanding, potentially opening new doors to research and technology innovations that can benefit humanity with direct societal relevance (NASEM, 2024b).

Urgent scientific research is time sensitive and has immediate societal relevance to emerging challenges at regional to global scales. Urgent scientific research needs to be done immediately in order to understand changes or new circumstances that can inform predictive models and decision-making and may be related to tipping point vulnerabilities (NASEM, 2024b).

Transdisciplinary research involves collaborators from multiple disciplines, knowledge systems, and sectors (including nonacademic) in a coequal partnership, collectively defining and developing new opportunities and ideas (e.g., conceptual understanding, theoretical models) toward tangible solutions that move beyond traditional disciplinary boundaries (Renn, 2021).

Note that transdisciplinary research is a distinct departure from *multidisciplinary research*, which involves several disciplines within academia whose contributions to a project are largely additive, rather than integrative. Transdisciplinary research is also distinct from *interdisciplinary research*, which involves partners from multiple disciplines in a potentially unequal relationship; it may not co-design a new future that spans disciplines. Interdisciplinary work provides a bridge between disparate disciplines, rather than co-creating knowledge that links to action. By contrast, transdisciplinary research strategies transgress and transcend sectoral boundaries, including strategies for identifying new and emerging problems and opportunities and for linking knowledge co-creation to problem-solving, thus providing needed information for policy decisions (Pohl, 2011; Russell et al., 2008; see Figure 1.2).

Forecast, Prediction, Projection are commonly used interchangeably. *Forecast* is a form of prediction. And in climate research, *projection* technically refers to a modelled outcome in the future based on assumed parameters or scenarios.

Scientifically, prediction is often used to describe likelihoods that are shorter term and more certain than forecasting, which tends to look at statistical probabilities, rather than certainties, that something will happen. For instance, in earthquake science, it is generally understood that prediction is either not possible (i.e., knowing ahead of time a specific time, place, and magnitude of an event) or not useful (e.g., saying there will be a magnitude 4 earthquake somewhere in the world tomorrow, which may be a true statement but is not particularly helpful). For example, an earthquake forecast might describe a 30 percent chance of a magnitude 7+ earthquake, with forecasted amounts of shaking along a known fault in the next 50 years. While an earthquake forecast may not be as immediately useful as a weather forecast (an 80 percent chance of rain tomorrow can tell you whether or not to take an umbrella), it can inform preparation such as construction standards and an early warning system, as well as insurance costs.

continued

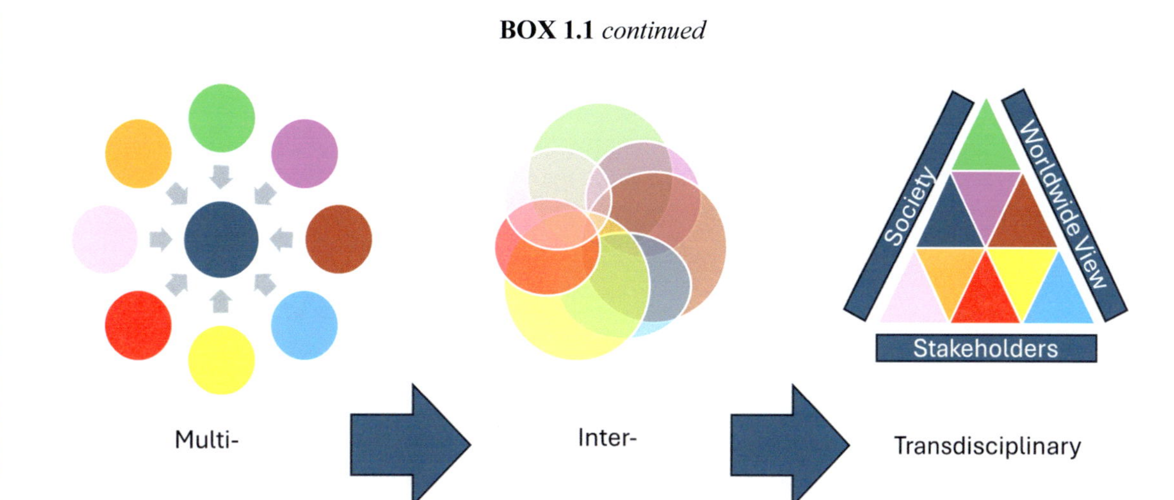

FIGURE 1.2 Graphic depiction of the difference between multi-, inter-, and transdisciplinary research.

This report generally uses the term *forecasting*, with the intent that vital basic and use-inspired research will lead to the ability to narrow the time windows and confidence of forecasts for extreme events at scales that matter for humankind, from earthquakes to extreme weather to rising sea levels and changes in ocean circulation.

The 2015 Decadal Survey also suggested a set of "Priority Science Questions for 2015–2025," eight high-level science questions that included topics that manifest at the ocean's surface, in water column processes, and on the seafloor. The research questions were broad, and research on each of them is ongoing; importantly, progress has been made not only by NSF but also through multiagency and international efforts and by the marine science community at large, and the questions remain relevant a decade later. A recommended best practice for the current decadal survey is to evaluate progress executed on these priority questions retrospectively. High-level examples of important progress made toward answering the 2015 Decadal Survey priority questions include the following:

- **Sea level change**: identified and clearly documented the acceleration of global sea level rise (Hamlington et al., 2024; IPCC, 2021) and contributions from ice sheet loss (The IMBIE Team et al., 2018, 2020; Wallis et al., 2023); elucidated factors that cause and/or contribute to regional sea level change, including atmospheric and oceanic processes (e.g., circulation) and land uplift and subsidence (Blackwell et al., 2020; Buzzanga et al., 2023; Dangendorf et al., 2024; Hamlington et al., 2020; Walker et al., 2023).
- **Ocean's role in modulating Earth's climate**: definitively calculated that the ocean has absorbed 90 percent of the excess heat produced by greenhouse gas warming (von Schuckmann et al., 2023) and 30 percent of the carbon dioxide (CO_2) released by fossil fuel emissions (Gruber et al., 2023), thus acting as a regulator to the climate known today; conclusively stated that 16 million years ago was the last time CO_2 concentrations in the atmosphere were consistently higher than today (The Cenozoic CO_2 Proxy Integration Project Consortium, 2023); developed an improved understanding of the connection between AMOC and heat and carbon uptake by the ocean (Caínzos et al., 2022; Fontela et al., 2016), indicating that the strength of AMOC is closely tied to possible tipping points of future circulation scenarios (Ritchie et al., 2021).
- **Importance of biodiversity**: linked the resilience of marine ecosystems to perturbations of their fundamental biodiversity (Duffy et al., 2016; O'Leary et al., 2017); scientifically described new species of ocean organisms across all taxonomic groups, made possible by advances in biotechnology tools (Jansson et al., 2023; Longo et al., 2022; Shin and Allmon, 2023; Teramura et al.,

2022), discovering that the deep biosphere holds a remarkable reservoir of genetic and taxonomic prokaryotic diversity on the planet (Bar-On et al., 2018); detailed how marine biodiversity affects marine ecosystem resilience and how both diversity and resilience are linked to patterns of marine food web structure.

- **Forecasting geohazards**: increased the ability to forecast earthquakes, fault slippage, and volcanic activity (Nooner and Chadwick, 2016; Obara and Kato, 2016; Panet et al., 2018; Ruiz et al., 2014; Socquet et al., 2017; Tan et al., 2016; Wilcock et al., 2016); improved predictability and early warning systems for tsunamis (Melgar and Hayes, 2019; Williamson et al., 2020).
- **Carbon sequestration in the deep ocean**: learned that microbial communities in the subsurface biosphere are playing a quantitatively relevant role in carbon removal (Kiel Reese et al., 2021; Shah Walter et al., 2018; Trembath-Reichert et al., 2021) and that the subseafloor ecosystem is home to the largest reservoir of microbial biodiversity and organic carbon on the planet (Jørgensen and Marshall, 2016).
- **Exploring the depths of the Earth**: drilled deeper into the Earth's mantle than ever before, while constraining conditions associated with the origin of life (IODP Leg 399 (McCaig et al., 2024).

Changing Times

The 2025 Decadal Survey is responding to a research landscape markedly different from that observed in the 2015 Decadal Survey. The balance between investments in research versus infrastructure still must be monitored and assessed strategically. But concerns have also risen around the balance between investments in basic research versus use-inspired or applied research. Underlying this concern is the increasing demand for applied research to resolve the global environmental problems society is facing.

NSF has shifted priorities beyond primary basic research to include use-inspired research with direct societal benefits. NSF's program Coastlines and People (CoPe) is an example of a use-inspired and interdisciplinary program that supports basic and applied research at the intersections between the natural, built, and social systems, with the goal of increasing resilience among the U.S. coastal populations and ecosystems. One important component of the CoPe program is expanding the science, technology, engineering, and mathematics (STEM) workforce. CoPe projects engage not only natural and social scientists but also community members to co-produce knowledge, and they weave impartiality into their design and implementation.

Further supporting use-inspired research and development, the Directorate for Technology, Innovation and Partnerships (TIP) was established in 2022 through the Creating Helpful Incentives to Produce Semiconductors and Science Act (known as the CHIPS Act) to advance U.S. competitiveness in STEM through investments that accelerate the development of key technologies and address pressing societal and economic challenges (NSF, 2024d). The TIP Directorate crosscuts the other six NSF research directorates to support co-design and co-development of multidisciplinary research aimed at translation from lab to market.

The Directorate of Geosciences (GEO), which houses OCE, has also made strides toward encouraging new partnerships both within and outside of NSF, as seen in the 2023 launch of the Division of Research, Innovation, Synergies and Education (RISE). As stated by NSF (n.d.-a), "The mission of RISE is to foster transdisciplinary collaborations that engage the broader geosciences community to drive transformative discoveries, innovations in workforce development, and use-inspired solutions for urgent Earth system challenges".[3]

While NSF and GEO have introduced new programs such as TIP and RISE, other programs have an uncertain future. Most notably, in 2023, OCE announced that the existing contract securing the workhorse of the U.S. scientific drilling program, the research vessel *JOIDES Resolution*, would not be renewed, thus ending the current phase of the program in 2024.

---

[3] See https://www.demo.nsf.gov/geo/rise/about.

In 2024, this committee released the 2025 Decadal Survey interim report, *Progress and Priorities in Ocean Drilling: In Search of Earth's Past and Future*, to provide advice on scientific research priorities and infrastructure needed to advance vital and urgent research priorities (NASEM, 2024b). The report also outlines what science questions can and cannot be addressed with existing infrastructure in the absence of a dedicated drilling vessel. The release of the report was followed by an announcement that the U.S. scientific ocean drilling program would transition to using a mission-specific platform model for at least the next 5 years. In July 2024, NSF announced the formation of a new subcommittee to outline the key infrastructure requirements of a new scientific ocean drilling platform that could meet requirements outlined by the scientific community while also meeting budgetary constraints (NSF, 2024c).

OCE is the only significant source of basic oceanographic research funding in the United States, and its focus of understanding fundamental ocean processes is necessary and needs to remain. This report highlights how basic and solutions-oriented science are beginning to intertwine; basic science understanding must underlie applied solutions in order for those solutions to be robust. This growing, intertwining relationship between basic and solutions-oriented science elevates the possibilities of integrating with other NSF programs and directorates, as well as other federal, state, and private funding entities.

Infrastructure and Workforce

Serious concerns impacting the effectiveness of ocean science research include the declining state of U.S. scientific infrastructure and maintaining a highly skilled U.S. workforce. Much of the critical infrastructure used in ocean science research, such as the ARF and the *JOIDES Resolution* (NASEM, 2024b), buoys, satellite systems, and marine laboratories need replacement or significant upgrades. The current replacement strategy is insufficient for satisfying today's needs and realizing scientific interests that align with environmental justice matters, including greater investment in capacity-building through constructing and maintaining research infrastructure (Ortner, 2013).

Additionally, a large portion of experienced ocean scientists and technical staff are retiring, working less, or need to adjust to new tools and technologies. For example, the commercial fishing community is experiencing a "graying of the fleet"—an increase in the average age of commercial fishermen, with fewer individuals entering the industry, leading to concerns about the future sustainability and resilience of fishing communities (Cramer et al., 2018; Haugen et al., 2021; Johnson and Mazur, 2018). Training for intergenerational scientific knowledge and skills-building, as well as training for new technology and tools, are needed for all career levels. This also includes training in collaborative processes and project management, as projects shift to become transdisciplinary in nature.

Technological Advancements

The last decade has also seen the significant development of new platforms and sensors and improved approaches to system integration to enable sampling not previously possible, although a need for coordinated and expanded sustainable ocean observations remains (NASEM, 2017a, 2020). Autonomous surface and underwater platforms are more widely available and are becoming effective tools for collecting a large amount of data in the ocean, expanding the footprint of traditional research vessels. While few of these developments can be credited to a single agency, NSF's role has been to catalyze the basic research needed for the initial development and deployment of technologies and enabling their transfer to mission agencies.

For example, while the technology for the roughly 3,900 Argo floats that now regularly sample the global ocean was developed in part with NSF funding, the current Argo program is largely funded by the National Oceanic and Atmospheric Administration (NOAA). Specifically, NSF has made a large investment to deploy over 500 Biogeochemical Argo (BGC-Argo) floats to supplement the Argo program (Matsumoto et al., 2022; NSF, 2019).

Ocean gliders are also regularly deployed across continental shelves, providing a snapshot of global ocean shelf conditions. One example of ocean glider deployment on a large scale is through a program called Boundary Ocean Observing Network, which remotely observes ocean environments, sharing data that are immediately relevant to coastal communities (Rudnick et al., 2017). Another example is the development of ecogenomic sensors on long-range gliders (Pargett et al., 2015).

New ocean sensors have been developed, such as distributed acoustic sensors on fiber optic cables; optical sensors that image and identify plankton and other organisms; and biogeochemical sensors that measure oxygen, nitrate, pH, and ocean alkalinity. The introduction of sensors such as passive acoustics (Yang et al., 2015) and biogeochemical sensors has also been funded through partnerships such as NOAA's National Oceanographic Partnership Program. These sensors provide critical (and difficult to sample) observations of the ocean's interior (Johnson et al., 2022).

The ability to integrate sensors into existing platforms has also increased significantly by commercialization and miniaturization of these sensors. It is now easier, faster and less expensive to add sensors to an existing platform or to build an integrated sensor array. Moreover, new approaches have been designed that allow scientists to partner with industry to leverage seafloor cables for seismic monitoring. Real-time communication and data storage costs have also been lowered, broadening access to agencies or communities utilizing the data for management actions. However, fundamental constraints remain, such as the limitations of data transmission underwater, the costs for real-time satellite communication and data transfer, and the large volumes of data exceeding the capacity of existing repositories.

Furthermore, over the last decade, the availability and usability of numerical models has increased significantly. Increased use of ocean and subseafloor models are the result of several key advancements, including (1) the availability of robust, well-understood physical models through accessible platforms, providing open-source code and a distributed version control system; (2) the development of model-sharing in key research areas such as climate and ice forecasting through model intercomparison projects; and (3) the integration of biogeochemical models, inclusion of human dimensions data, and, to a lesser extent, coupling of biological models into physical models. These models and model results are now used by researchers regularly at many different stages of their research program. And many of these models are now being adapted to take advantage of widely available computing infrastructure, such as cloud and GPU computing. The effort now required to access or apply a model has been substantially reduced thanks to these advances. The use of numerical models to help answer robust scientific research questions has thus become more prevalent, enabling researchers to extract meaning from their data faster and more effectively, and to understand where gaps in ocean data occur (Graw et al., 2021).

The availability of models to run routine forecasts and hindcasts of ocean conditions has also increased significantly. For example, new machine learning methods, such as applying AI to model forecasting, are still in their infancy but offer promise for studying geophysical processes and making accurate ocean "weather" forecasts (Price et al., 2025). It is important to note that the numerical models are still far from perfect, and the need exists for improving understanding of dynamics within the models, specifically the coupling of biological/biogeochemical processes with physical processes to forecast with any accuracy the response of ocean systems to global change. Basic ocean science research, such as targeted and sustained ocean observations, are sorely needed to inform and improve the models and thus the forecast.

Additionally, because of data limitations, many machine learning models require high-fidelity simulations as training sets. The effect of inaccuracies in simulated training data is not well-understood, driving the need to improve the underlying models. High-resolution geologic records from the subseafloor, which provide paleo-analogs for possible future ocean and climate states, are also needed to inform and improve the models and thus the forecasts. In addition, geophysical observations are required to test new generations of models for earthquake ruptures, landslide generation, and volcanic eruptions.

Societal Shifts

Over the last decade, societal developments have changed within the ocean science landscape. This led to initiatives and efforts at sector levels to bring forth the importance and benefits of a more multifaceted approach to ocean sciences, ensuring that all people have opportunities to have access to, contribute to, and benefit from ocean research. This approach acknowledges that access to ocean science has been limited, and it seeks to expand engagement beyond academia through knowledge-sharing and capacity-strengthening. Such initiatives aim to realize scientific excellence from more perspectives, strengthening its rigor and societal relevance. These concepts were not addressed in the last decadal survey but have been discussed throughout the deliberations of the current committee.

Coastal counties are often more demographically heterogeneous than noncoastal counties (U.S. Census Bureau, 2019). However, the scientific community does not generally resemble the demographics of coastal communities (Harris et al., 2022). This mismatch may limit the value of research if is not relevant to the communities or if the scientific findings are not effectively communicated (Behl et al., 2021; Harris et al., 2022). Increasing participation of those underrepresented sectors in the ocean science enterprise contributes to sustaining the nation's capacity for innovation and discovery. These underutilized constituencies have great potential to improve outcomes in both basic and applied sciences.

Need for Renewed Leadership in Ocean Research

The decade covered by the 2015 Decadal Survey ended on a challenging note, with budget cuts to NSF and significant budget cuts to OCE announced. The loss of the scientific drillship and the mismatch between the delivery of new research vessels and retirement of the old reduced the United States's capacity to send scientists out to sea. These are instructive lessons on the need for long-term planning for the future of ocean science, especially as it involves specialized infrastructure. Now is the time for the United States to invest and take leadership in the ocean science priorities required for a healthy, predicted, and understood ocean, which is critical for national security and economic prosperity in the next decade and beyond.

2025–2035 Decadal Survey of Ocean Sciences

As the next decade (2025–2035) approached, NSF asked the National Academies to develop a new decadal survey to provide guidance to the agency on research and infrastructure strategies to address priority research questions surrounding ocean and Earth system science. Specifically, NSF sought advice on how to incorporate use-inspired research into its 2025–2035 research portfolio in a way that complements the traditional emphasis on basic research. In so doing, it recognized that strategies for the next decade need to take advantage of new technological advances and capabilities, as well as strategic partnerships.

Additionally, the committee was tasked with advising NSF on potential changes in workforce training and support needed to address the new research priorities, such as building transdisciplinary teams and strategies for using scientific information to help address societal challenges. This includes considering goals for broadening access and participation with an embracive lens in the ocean sciences and how progress toward these goals should be measured and reported.

The 2025 Decadal Survey was developed to ensure that OCE will continue to provide the foundation for ocean science contributions, not only within NSF but also across the federal agencies that use ocean science to accomplish their missions. The committee's statement of task is provided in Box 1.2.

STUDY STRATEGY

The committee met with representatives from NSF to ensure its interpretation of the statement of task (Box 1.2) aligns with sponsor expectations. The committee's interpretation of the task is that OCE will continue to fund basic research across the ocean sciences, as it has always done. The purpose of the 2025 Decadal Survey is to highlight a few (2–3) use-inspired and societally relevant research priorities that OCE—together with other NSF units and with agencies and organizations outside NSF—could also invest in to address urgent questions surrounding ocean and Earth system science. The research priorities put forward in this report are intended to add to, and not replace, continued funds for basic research. Additionally, OCE is not looking to the 2025 Decadal Survey for advice on budget balancing nor specifics on how to accomplish the research agenda. Rather, the committee was tasked with identifying a few focus areas and the infrastructure (both physical and human resources) needed to advance U.S. ocean science research by 2035.

BOX 1.2
Statement of Task

The Decadal Survey will advise the National Science Foundation's Division of Ocean Sciences (NSF OCE) on forward-looking approaches to guide investments in research, infrastructure, and workforce development. The committee will develop a compelling research and infrastructure strategy to advance understanding of the ocean's role in the Earth system and the sustainable blue economy. The report will recommend ways that NSF OCE could develop the capacity to respond nimbly as priorities change and new opportunities emerge over the 2025-2035 decade.

The committee will produce an interim report to provide advice to NSF OCE on the resources and infrastructure available to address high priority research questions requiring scientific ocean drilling. The interim report will cover the following:

1. Based on previous reports, assess progress on addressing high priority science questions that require scientific ocean drilling and identify new, if any, equally compelling science questions that would also require scientific ocean drilling.
2. Of the unanswered scientific questions, which could be addressed through the use of existing scientific drilling assets including sediment or rock core archives and existing platforms, and which questions would require new infrastructure or sampling investments?

The final report will address the following:

1. Identify novel opportunities regarding ocean-related, use-inspired, solutions-oriented research and innovation. This assessment will include specific examples of opportunities for the Division to make substantial contributions to and develop collaborative and complementary research efforts with NSF's Directorate for Technology, Innovation, and Partnerships (TIP).
2. Identify opportunities and strategies to promote innovative multidisciplinary and multi-sectoral approaches to address complex science challenges arising from the intersection of natural processes, societal needs, and human-driven environmental change. This will include strategies for training the next generation of ocean scientists and incorporating the principles of diversity, equity, inclusion, environmental justice, and access into these scientific endeavors.
3. Develop a concise portfolio of compelling, high-priority, scientific questions that have the potential to transform scientific knowledge of the ocean and the critical role of the ocean in the Earth system. Identification of the scientific questions will update the priorities identified in Sea Change: Decadal Survey of Ocean Sciences 2015-2025, drawing from recent reports and community input, including recent National Academies reports and activities such as the U.S. National Committee for the UN Decade of Ocean Science for Sustainable Development. The selection may be based on timeliness, societal benefits, technological advances, or other criteria as identified by the committee.
4. Identify the research infrastructure needed to advance the high-priority ocean science research questions identified in Task # 3, including an assessment of current facilities and the potential for future investments and development of new technologies to meet the needs of the research community. The assessment will include the committee's perspectives on the relative need for continued funding of specific infrastructure and mechanisms to evaluate the contributions of major infrastructure to the research enterprise.
5. Develop a framework that OCE can apply to leverage and complement the capabilities, expertise, and strategic plans of its partners (other NSF units, federal agencies, private sector – such as ocean industries and foundations, and international organizations). The framework will include approaches to encourage greater collaboration and maximize shared use of research assets and data.

In undertaking these tasks, the committee will engage the ocean science community and other relevant fields to gather ideas and develop recommendations informed by broader community perspectives. The final report will include assessment of challenges and identification of metrics for progress in achieving the vision of the decadal survey. The 2025-2035 Decadal Survey committee will address these tasks within the context of the current OCE budget while identifying aspirational goals that NSF could implement with growth in the OCE budget over the decade.

Convening and Information-Gathering

To complete this study, the project included convening an ad hoc committee of 23 ocean science researchers and practitioners, representing a wide range of discipline, institution, sector, geography, age, race, and ethnicity (their biographies are included in Appendix A). The committee's expertise was augmented by additional invited professionals to join the information-gathering sessions described in the following section. Hybrid committee meetings were held over the course of roughly a year for information-gathering and closed committee deliberations, as well as monthly virtual committee meetings and various topical subcommittee meetings.

From June 2023 to July 2024, the committee held six 2-day hybrid meetings at various locations around the United States, and monthly 2-hour full committee meetings to gather information, deliberate, and write and publish the interim and final reports. During these meetings, 101 experts were invited to speak with the committee on various topics. The information-gathering portions of all meetings were open to the public and recorded and archived on the project's website.[4] Appendix B includes agendas for all public meetings. Through these public meetings, the committee heard arguments for future research broadly needed across the physical, chemical, biological, and marine geological and geophysical sciences, including from experts on applied and multidisciplinary topics, such as scientific ocean drilling, effects of a changing climate, marine carbon dioxide removal, ocean acidification, marine biodiversity, marine critical minerals, and urban seas. The committee also heard from those on the infrastructure side of ocean sciences to understand future anticipated needs and capabilities in support of ocean science research, such as the ARF, marine Long-Term Ecological Research sites, the Ocean Observatory Initiative, the Argo and BGC-Argo programs, and scientific ocean drilling. Throughout the information-convening process, discussions were held on programs that have benefited greatly from co-developed and co-produced research and that have moved toward a more transdisciplinary way of conducting science. In addition to extensive information-gathering meetings, the committee solicited input on research priorities from the broad ocean science community. The committee hosted two virtual town halls, receiving 117 responses.[5] The committee also held in-person town halls at the 2023 American Geophysical Union meeting and the 2024 Ocean Sciences Meeting.

Interim Report

In accordance with the statement of task, the first published result of this study was an interim report—*Progress and Priorities in Ocean Drilling: In Search of Earth's Past and Future* (NASEM, 2024b), which focuses on the future of scientific ocean drilling—while the full report was still being developed. This interim report, released to the public in March of 2024, covered research priorities that can be progressed only through scientific ocean drilling and the infrastructure needed to accomplish such research. The conclusions from the interim report are folded into the prioritization of research and infrastructure discussed in this report and considered in the context of all ocean sciences.

IDENTIFYING AND PRIORITZING OCEAN SCIENCE RESEARCH NEEDS

In addition to considering input gleaned from its extensive information-gathering sessions, town halls, and unsolicited emails or other communications, as described above, four different committee-wide exercises (held over the year) helped bring all the ideas for future ocean science research and infrastructure to the surface for evaluation. The committee's deliberations included review of the following (in no particular order):

---

[4] See https://www.nationalacademies.org/our-work/2025-2035-decadal-survey-of-ocean-sciences-for-the-national-science-foundation.

[5] See https://app.smartsheet.com/b/publish?EQBCT=1cd55d1f6f0a4abebd14d0724b76670e (accessed September 30, 2024).

- Recent relevant federally issued reports, such as *Science and Technology for America's Oceans: A Decadal Vision* (SOST, 2018); *Opportunities and Actions for Ocean Science and Technology, 2022–2028* (SOST, 2022); *Ocean Climate Action Plan* (OPC, 2023a); *The United States Ocean Acidification Action Plan* (NOAA, 2023); *Ocean Justice Strategy* (OPC, 2023b); *National Strategy for a Sustainable Ocean Economy* (OPC, 2024); and *The National Ocean Biodiversity Strategy* (SOST, 2024).
- Relevant National Academies reports, such as *Sea Change: 2015–2025 Decadal Survey of Ocean Sciences* (NRC, 2015); *Earth System Predictability Research and Development: Proceedings of a Workshop—in Brief* (NASEM, 2020); *Cross-Cutting Themes for U.S. Contributions to the UN Ocean Decade* (NASEM, 2022a); *Next Generation Earth Systems Science at the National Science Foundation* (NASEM, 2022b); *A Research Strategy for Ocean-Based Carbon Dioxide Removal and Sequestration* (NASEM, 2022c); *Future Directions for Southern Ocean and Antarctic Nearshore and Coastal Research* (NASEM, 2024d); and *Progress and Priorities in Ocean Drilling: In Search of Earth's Past and Future* (NASEM, 2024b).
- Review of resources from the Intergovernmental Panel on Climate Change and associated updates (IPCC, 2023).

The committee compiled, analyzed, and discussed in detail research needs across the field of ocean sciences during closed committee meetings and evaluated research needs based on the criteria included in Box 1.3. Research topics meeting these criteria were elevated for consideration by the committee as a priority for the next decade.

BOX 1.3
Evaluation Criteria for the 2025 Decadal Survey Research Priorities

Appropriateness for the unique responsibilities of the National Science Foundation (NSF) and its Division of Ocean Sciences:
- ✓ Does the research require advances in basic understanding?
- ✓ Does the research fall inside the bounds of the NSF research realm?
- ✓ Can this research be classified as use-inspired, solutions-oriented research and create transferrable knowledge?

Importance:
- ✓ Is the research compelling and timely, innovative and/or groundbreaking?
- ✓ Does the research lead to an understanding of the ocean and ocean-adjacent research?
- ✓ Would the research yield critical knowledge that will lead to new avenues or approaches of research, or unlock other critical research pathways?

Impact:
- ✓ Would the research impact the wider scientific community?
- ✓ Does success in addressing the research result in U.S. leadership in advancing multifaceted research, education, and workforce development?
- ✓ Would the research address grand challenges faced by humanity relating to global-scale societal, economic, and ecological change?

Transdisciplinarity (defined in Box 1.1):
- ✓ Does the research promote progress at the intersection of disciplines and across societal sectors?
- ✓ Will the research benefit from, or require, collaborations and coordination across disciplines and multiple knowledge systems?

REPORT STRUCTURE

The report is organized into five chapters. This chapter has provided the study context and background and establishes the critical role of ocean sciences to human well-being. Chapter 2 presents the committee's assessment of ocean science research priorities most urgent for NSF and other organizations to support over the next decade. Chapter 3 discusses the necessity of innovating approaches in order to accomplish the research priorities. Chapter 4 discusses the infrastructure required and relevant considerations, including investment needs in a broader sense to facilitate U.S. leadership in basic ocean sciences in the next decade and beyond. Chapter 5 pulls the report content together in the form of recommendations for NSF by presenting different funding scenarios for helping the United States and the world better observe, understand, and forecast our future. Doing so will enable the research community to provide necessary information to communities and scientists to facilitate the adaptation, resiliency, and prosperity of humankind and the ecosystems foundational for life and well-being.

2
Urgent Ocean Science Research Portfolio

In his report that led to the creation of the National Science Foundation (NSF), Vannevar Bush (1945) argued for the importance of basic research, citing the "flow of new scientific knowledge" and the understanding that this "new knowledge can be obtained only through basic scientific research" (pp. 1, 7). Bush argued strongly for U.S. leadership in the pursuit of science; regaining such leadership is still important today, not only for national security but also for the United States's ability to attract and convene the best scientific minds, generating a workforce with skills that will fuel a thriving economy.

Confirming Bush's vision, basic research in the field of ocean sciences has resulted in advances in predictive technologies that have led to benefits for humans, including reductions in loss of life. As one example, basic ocean research on air–sea interaction and surface wave dynamics has dramatically improved path and intensity forecasts for hurricanes and their resulting storm surges, thereby saving lives and reducing economic loss (Cangialosi et al., 2020; Miles et al., 2021). The ability to predict or forecast other ocean processes is essential for adapting to change and mitigating loss (Link et al., 2023). For example, basic research has revealed that ocean species can impact the physical and chemical nature of the ocean, including aiding in carbon sequestration, absorption of heat in the surface ocean, influencing the flow of nutrients, and even impacting wave energy. Understanding how ocean life populations adapt to environmental change is essential for forecasting ecosystem resilience.

This chapter emphasizes the importance of and need for basic research and shares urgent ocean research priorities, organized in three overarching themes and framed under a single challenge for the next decade.

BASIC RESEARCH IS KEY

The program officers at the Division of Ocean Sciences (OCE), as those in other divisions of NSF, act as fair brokers to evaluate ocean science research proposals; after conferring with external peer reviewers and a panel or committee of experts, program officers decide which proposals to fund. The goal of the process is to fund those proposals that best stand to advance understanding of the ocean and its role in the Earth system and that have impacts on education and other societal benefits in addition to the proposed research objectives. Funded proposals are often in the category of basic research that lacks immediate known applications, although use-inspired proposals are also encouraged. And in some cases, NSF research announcements solicit proposals that encourage research related to a specific societal benefit. However, it is the support of *unsolicited* proposals for basic ocean science research that distinguishes OCE from other federal mission agencies in the ocean sciences. This basic research underpins the eventual applied societal benefits.

Over the course of this study, the committee evaluated testimony from the scientific community on the future basic research needed across the field of ocean sciences, including applied and multidisciplinary topics, such as effects on ecosystems, marine geoengineering strategies, potential benefits and hazards associated with a new economy for marine critical minerals and carbon sequestration, precursors to subduction zone geohazards, changes in biodiversity and biogeography patterns for marine organisms, and the multistressed urban seas and coastal ocean (see Appendixes B and C).

CONCLUSION 2.1: NSF is the only U.S. federal funding agency intentionally founded with the explicit mission to sponsor basic research. Such supported research produces the raw intellectual knowledge that will yield tomorrow's solutions. To support national security, a skilled U.S. workforce, and a thriving economy, continued, dedicated NSF funding is critical for principal investigator–driven research projects that address questions fundamental to understanding the ocean and its interactions with other parts of the Earth system.

RECOMMENDATION 2.1: The National Science Foundation's Division of Ocean Sciences should continue to support a broad portfolio of basic research to ensure that scientists and engineers of the future have access to a continually improved understanding of the ocean to advance and fuel innovation and resilience in the United States.

A CHALLENGE FOR THE NEXT DECADE

Based on the prioritization work described in Chapter 1, the committee developed an overarching challenge for the next decade of ocean sciences: forecasting the state of the ocean at scales relevant to human well-being (Box 2.1).

BOX 2.1
The Challenge

By 2035, establish a new paradigm for forecasting ocean processes at scales relevant to human well-being, emphasizing three interconnected themes: ocean and climate, ecosystem resilience, and extreme events. Establishing this paradigm will require a relentless focus on furthering basic science to observe and understand ocean processes through the lens of innovative disciplinary, as well as transdisciplinary, research practices.

The committee elevates the ability to forecast as its challenge to emphasize the importance of developing testable hypotheses of future conditions. Figure 2.1 illustrates the steps leading to a forecast, as well as the feedbacks from initial forecasts that provide priorities for improving measurements and for better understanding of the process(es) under study. While forecasting is elevated here, all steps in the continuum are essential. Better understanding will in turn improve new forecasts. Forecasts are also key to addressing use-inspired issues beyond basic research and for identifying and prioritizing efforts that are needed to respond to changes in the Earth system that affect human well-being.

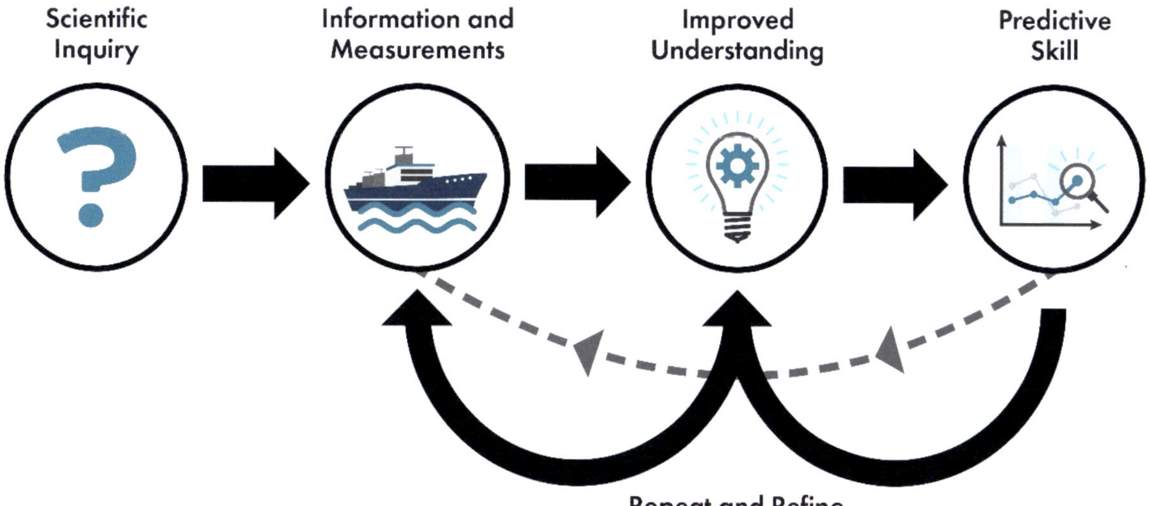

FIGURE 2.1 Iterative process of forecasting the ocean state.

Translating basic research into forecasts that are operational requires partnerships with other mission agencies focused on ocean sciences. For some challenges (such as some extreme events), society might best be served with forecasting timescales of days to weeks; others (such as sea level rise) may require longer timescales. Similarly, decisions on fishery management, ecosystem services, or renewable energy considerations potentially contain multiple decision time- and space scales, so no single forecasting system can be constructed that can be helpful in all cases. In other words, forecasting the state of the ocean at scales relevant to human life must occur at various temporal and spatial scales as dictated by the societal questions at hand and by the nature (e.g., rates of change, response times) of the ocean system processes.

Research is also needed on the limits of the predictability of the ocean system. These limitations may be due to the lack of detailed knowledge of the state of the system or to inherent complexities in the system that may lead to chaotic behavior. Thus, some ocean processes may be difficult, if not impossible, to predict with any accuracy (NASEM, 2020). These complexities may necessitate the development of probabilistic approaches to eventually serve as components of forecasting systems.

The ultimate goal of forecasts is to contribute to federal management and policy decisions by government and private entities that promote resiliency and prosperity, and position the United States as a leader in transformative ocean science. OCE-sponsored basic research, including transdisciplinary research, is necessary to achieve this goal of operational forecasting for as many processes as possible.[1]

2025–2035 OCEAN RESEARCH PORTFOLIO

Basic research across all disciplines of ocean sciences is sorely needed, and an argument can be made for elevating the importance of almost any topic. Yet the committee's task (Box 1.2) was to develop a "concise portfolio of compelling, high-priority, scientific questions that have the potential to transform scientific knowledge of the ocean and the critical role of the ocean in the Earth system." Based on criteria included in Box 1.3, and in pursuit of the challenge (Box 2.1), the committee prioritized the following urgent, high-priority science themes and questions for the next decade—ocean and climate, ecosystem resilience, and extreme events (Figure 2.2).

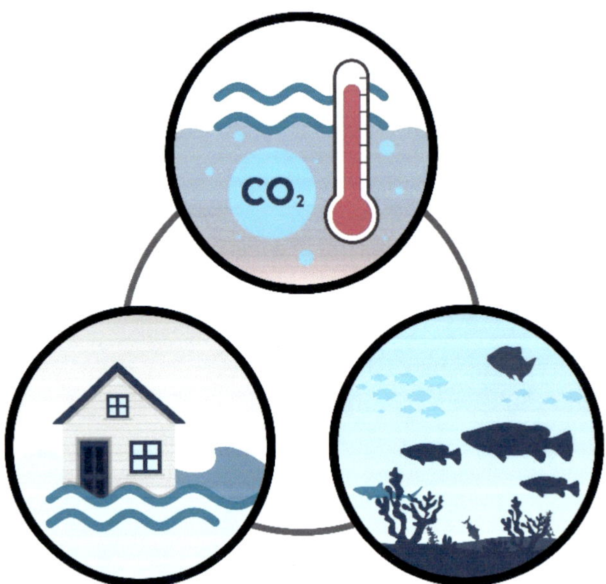

FIGURE 2.2 Interconnectivity and integration of the three 2025 Decadal Survey priority research themes.

---

[1] *Transdisciplinary research* is defined in Box 1.1 in Chapter 1 and further discussed in Chapter 3.

This compelling, cross-cutting research portfolio should galvanize the ocean science community to work together towards the challenge, and common goal, of establishing a new paradigm for forecasting ocean processes at scales relevant to human well-being by 2035.

CONCLUSION 2.2: In addition to continued NSF funding for wide-ranging basic ocean science research, three themes and research questions urgently need to be addressed through the application of improved forecasting capabilities in the next decade and beyond:
- *Ocean and climate—How will the ocean's ability to absorb heat and carbon change?*
- *Ecosystem resilience—How will marine ecosystems respond to changes in the Earth system?*
- *Extreme events—How can the ability to forecast extreme events driven by ocean and seafloor processes be improved?*

RECOMMENDATION 2.2: The National Science Foundation's Division of Ocean Sciences should support basic research that addresses the goal of forecasting ocean processes at scales relevant to human well-being, with emphasis on the following themes and questions:
- **Ocean and Climate—How will the ocean's ability to absorb heat and carbon change?**
- **Ecosystem Resilience—How will marine ecosystems respond to changes in the Earth system?**
- **Extreme Events—How can the ability to forecast extreme events driven by ocean and seafloor processes be improved?**

Each of these themes and questions are explained further in the sections that follow. In many respects, they amplify the priorities highlighted in past reports (e.g., NASEM, 2017a, 2022a, 2022b, 2024b, 2024c; NRC, 2015), and thus they contain lines of inquiry that are familiar and well supported. Here, however, they are combined in ways that emphasize the connections between elements of ocean processes and ocean life that are fundamental to accurate dynamic forecasting.

The priority research themes are both goal oriented and purposefully general, providing space for researchers to connect their research to the theme while still advancing towards a common goal. Table 2.1 provides an overview of the types of research questions that could be included within each theme, as well as the potential societal uses and outcomes of the research and common resources needed to advance all three themes, with context following in the remainder of the chapter. Each of the next three subsections begins with the importance and urgency of each theme and continues with background and context. The societal relevance and broader impacts of the research are also highlighted, followed by a high-level discussion on the resources required for successful conduct of such research.

TABLE 2.1 Urgent Priorities for Ocean Sciences Research, 2025–2035

OCEAN AND CLIMATE	
Urgent Question	**How will the ocean's ability to absorb heat and carbon change?**
Example Research Directions	• Developing new approaches to observe Atlantic meridional overturning circulation (AMOC) strength, specifically heat transport in the North and South Atlantic • Improving predictions of marine ice sheet instability driven by warming ocean waters • Obtaining better observations of surface partial pressure of carbon dioxide (CO_2) in the Southern Ocean in the winter to constrain ocean CO_2 uptake • Developing ways to help quantify variability in the carbon cycle (e.g., with new biogeochemical observations and sensors) • Testing and validation of models with observations of past changes in the carbon cycle and climate (e.g., mid-Miocene climate optimum, mid-Pliocene warm period, and rapid glacial to interglacial transitions)

continued

TABLE 2.1 *continued*

	OCEAN AND CLIMATE

Example Use-Inspired Cases	• Marine carbon dioxide removal (mCDR): Understand potential for the ocean's natural CO_2 removal and sequestration processes to be accelerated safely and at a scale that would draw additional CO_2 from the atmosphere and mitigate changes in ocean pH
	   o Develop technology and techniques to evaluate the effectiveness of different approaches to mCDR
	   o Evaluate how changes to marine ecosystems affect the transfer of carbon to the deep ocean
	   o Evaluate the suitability of subseafloor geologic formations for carbon storage
	• Identify potential tipping points in the Earth system:
	   o Forecasting if or when AMOC reaches the point of irreversible change
	   o Ice sheet collapse

Potential Outcomes	• Improved forecasts of potential tipping points • Ability to monitor and predict ocean uptake of CO_2 • Better understanding of changes to the processes moving carbon from the atmosphere to deep ocean waters
	ECOSYSTEM RESILIENCE
Urgent Question	**How will marine ecosystems respond to the changing Earth system?**
Example Research Directions	• Determining the effects of warming, acidification, and deoxygenation on the spatial and temporal dynamics of ocean populations, from local (human) to biome scales, in order to anticipate the early signals and downstream consequences for marine ecosystems • Predicting the weakening of the biological pump/export production and how this will affect trophic structure and function and, subsequently, food security, and exploring solutions for integrating multiple knowledge systems and resource management approaches • Deciphering practical, useful understanding, forecasts, and decision criteria pertaining to the resilience of open-ocean, coastal, and estuarine ecosystems • Codeveloping tools for rapid measurements and assessments of ocean biological and functional diversity—specifically using advancements in environmental and organismal DNA, as well as omics sciences • Understanding and forecasting energy and biomass transfer in and across marine ecosystems and how that transfer contributes to resilience • Understanding the effects of seafloor mining or deep-sea ecosystems and what their response and recovery may be, and respective timescales
Example Use-Inspired Cases	• Sustainable fisheries: Forecast impacts of ocean acidification, warming temperatures, and sea level rise, as well as other demands from multiple ocean-use sectors (including overfishing) on food web dynamics essential to food security • Storms or sea level rise: Understand the baseline seasonal variability in plankton dynamics in coastal or coral reef ecosystems, then examine shifts in biodiversity and timescales for resilience during storms, sea level rise, or related extreme events • Habitats and coastal biostructure: Understand how future ocean changes will impact the ability of foundational ecosystems—such as coral reefs, coastal marshes, and mangroves—to support human communities
Potential Outcomes	• Better understanding of organisms at the base of the food web, the trophic links and biomass transfers, and models, ultimately predicting changes in food security • Development of interventions to support recovery of damaged coral reef and other coastal ecosystems • Improve models that link short-term changes to ocean processes with impacts on fisheries • Spatial forecasts of ocean changes to support creation and adaptive management of effective area-based conservation measures
	EXTREME EVENTS
Urgent Question	**How can the ability to forecast extreme events driven by ocean and seafloor processes be improved?**
Example Research Directions	• Integrating existing and new data and knowledge to elucidate precursory signals, if they exist, that can improve forecasting of earthquakes, underwater volcanic eruptions, submarine landslides, and associated extreme events (e.g., tsunamis), and thus improve early warning systems • Increasing ability to model the future impact of ocean processes on global weather extremes; forecasting the extent of deoxygenation and dead zones and thus anticipating the impact on ecosystem health and fisheries

	• Producing extreme event forecasts to inform new societal adaptation and ecosystem restoration approaches that use nature-based design features • Enhancing forecasts by integrating existing and new data and knowledge from across disciplines and communities to improve responses to extreme events
Example Use-Inspired Cases	• Coastal resilience: Improve co-development of mitigation strategies with coastal communities that are potentially impacted by extreme events such as coastal flooding • Sustainable agriculture: Increase efficiency in natural resource production through accurate forecasts that inform sustainable agricultural and forestry practices • Infrastructure investment: Improve forecasts of ocean extreme events from sea level rise to earthquakes to provide guidance on sustainable urban planning, safety, and insurance • Instrumentation of existing infrastructure, such as telecommunication cables for monitoring: Partner with industry to provide real-time offshore seismic monitoring or carry out temperature-sensing to track ecosystem health for blue economic development
Potential Outcomes	• Improved crop yield due to ability to plant with long-term forecasts in mind • Transdisciplinary research to improve the safety of communities vulnerable to extreme weather events and geophysical hazards • Sustainable fisheries and healthy ecosystems in waters prone to oxygen depletion
CLIMATE AND OCEAN; ECOSYSTEM RESILIENCE; EXTREME EVENTS	
Resources Needed	• Ocean observing—Global persistent observations, including ocean's interior using strategic and creative partnerships to meet observational and technological needs; continued support of long-term observational programs, such as the Bermuda Atlantic Time-series Study, the Hawaii Ocean Time-series, Long Term Ecological Research sites, Science Research Centers, the Global Ocean Ship-based Hydrographic Investigations Program, and the Argo program • High-resolution models validated by observational data • Palaeoceanographic records • Broadened workforce • Strategic and creative partnerships to meet observational and technological needs • Expanded data curation and integration • Improved data assimilation, machine learning, or hybrid approaches for comprehensive model calibration and initialization • Next-generation computer architectures for advanced simulations and big data analyses

OCEAN AND CLIMATE

How will the ocean's ability to absorb heat and carbon change?

The ocean is presently absorbing 90 percent of the heat (von Schuckmann et al., 2023) and roughly 30 percent of the carbon that result from global emissions of greenhouse gases (Gruber et al., 2023). Any decline in these rates of uptake would accelerate increases in CO_2 levels and temperature in the atmosphere. In essence, the ocean has been operating to slow the impacts of climate change felt on Earth since the industrial revolution. It is unlikely, however, that the ocean will continue this same rate of absorption (Wilson et al., 2022), and it is likely that the patterns of ocean current circulation will shift; these patterns are crucial for carbon and heat distribution (Gray, 2024; Gruber et al., 2023; Liu et al., 2022). Scientists do not know whether the expected changes will be gradual or if there is a threshold where such shifts will reach a *tipping point*, where the shift becomes sudden, dramatic, and potentially irreversible. Shifts in how the ocean absorbs heat and carbon can have differing and far-reaching impacts on global ocean circulation and atmospheric processes that include hurricane and storm development and movement; ice sheet stability; ocean chemistry, including acidification, nutrient speciation, and cycling; and ocean productivity and ecosystem health, including major die-off events (e.g., coral bleaching). Research has uncovered the ways in which the Earth system is coupled with the ocean and other systems, yet there is limited understanding about how the coupled relationships will change in the future. The pace of change in our climate, in particular, linked to heat and carbon stored in the ocean, creates an urgency to understand the changes Earth might experience in the coming decades.

Background and Context

Through its capacity to store and transport carbon and heat, the ocean plays a key role in regulating the climate and the carbon cycle (Box 2.2). Observations and models reveal that the ocean has played a critical role in regulating current and past changes in climate and atmospheric carbon concentrations (e.g., glacial–interglacial cycles). For example, studies of cored subseafloor sediments recovered by scientific ocean drilling have demonstrated a connection between AMOC, regional climate change, and carbon uptake by the ocean. This relationship indicates that changes to the strength of AMOC could be a key component of the ocean's response to future warming, including altering ocean circulation patterns, which would impact global heat transport, potentially affecting regional climates, global atmospheric circulation, and the hydroclimate in general (e.g., Ait Brahim et al., 2022; Sallée et al., 2023; Walczak et al., 2020; Wu et al., 2021).

Long-term, sustained observations supported by OCE in partnership with other directorates and agencies have provided some of the most critical insights into how the ocean's physical and chemical properties have changed over the last several decades. This includes long-term time-series sites, such as Station ALOHA (sampled by the Hawaii Ocean Time-series) and the Bermuda Atlantic Time-series Study; long-term regional programs, such as the California Cooperative Oceanic Fisheries Investigations survey, coastal and ocean Long-Term Ecological Research sites, the global Argo float program; and the Global Ocean Ship-based Hydrographic Investigations Program (GO-SHIP), which has resampled ocean transects that were established in the 1990s. GO-SHIP has provided crucial ship-based measurements on global ocean carbon and related parameters, unavailable to any other observation platform, and has enabled an accounting of the ocean anthropogenic CO_2 sink. Higher-frequency regional and seasonal observations by biogeochemical (BGC)-Argo have been needed to complement shipboard measurements, which in turn provide calibration data for Argo, to study the Southern Ocean's role in CO_2 uptake (Bushinsky et al., 2019).

Understanding the Ocean Carbon Sink

Researchers are just beginning to understand the role of the Southern Ocean in CO_2 uptake (Gruber et al., 2019; Terhaar et al., 2021), as austral winter measurements become more prevalent (Bushinsky et al., 2019). These observations point to the importance of local, seasonal (both winter and summer), and sub-seasonal episodic events (e.g., storms)—which are much more effectively captured by floats and other sustained measurements—for accurately estimating CO_2 uptake in the Southern Ocean (Carranza et al., 2024; Hauck et al., 2023). Recent studies have also demonstrated the importance of considering vertical mixing alongside subpolar gyre circulation and the entrainment of carbon in the horizontal circulation, zonal asymmetric in Antarctic circumpolar current, as well and air–sea exchange (Gray, 2024), for CO_2 uptake. This contrasts with previous studies that have suggested a central role for meridional overturning circulation in controlling the air–sea CO_2 flux and CO_2 inventory in the ocean's interior (MacGilchrist et al., 2019).

BOX 2.2
The Ocean Carbon Cycle

The carbon content of the ocean increases from the surface to the deep sea. The ocean is stratified with different water masses layered horizontally: a relatively thin, low-density, warm layer overlays layers of thicker, denser, colder, and deeper waters. This layered structure limits exchange of carbon between the near-surface waters and the deep layer that holds most of the carbon. Exchange of carbon between these layers occurs through both physical and biological exchanges (Figure 2.3). The *physical exchange* delivering carbon to the deep ocean occurs mainly in the polar and subpolar regions. Traditionally, the *biological exchange* has been characterized as dead marine life sinking from the surface to deeper waters, where much of the carbon is respired as this organic matter decays. This is often referred to as the export flux or the biological carbon pump (BCP). Although the traditional view of the BCP implicates gravitational settling, new observations have determined that a number of other pathways—such as subduction zone disturbance, mixing by wind and currents, and zooplankton migration up and down the water column—may be as important as the gravitational BCP in exporting carbon from the upper ocean in certain regions (Boyd et al., 2019; Nowicki et al., 2022; Omand et al., 2015; Stukel et al., 2017). Predicting how the relative importance of the different pathways will change in the future is challenging, especially as marine ecosystems, and thus human communities, change. Without the BCP, atmospheric carbon concentrations would be roughly 200 parts per million (ppm), or about 50 percent higher (Henson et al., 2022). The respired carbon that accumulates in the deep ocean is eventually returned to the surface after hundreds of years through physical mixing.

Given the mass of the ocean carbon reservoir, the ocean will play a key role in moderating carbon levels in the atmosphere. Over the last century, more than 30 percent of the carbon emitted to the atmosphere has been absorbed by the ocean (Gruber et al., 2019). In the future, as the ocean surface continues to warm, the rate at which the surface absorbs carbon from the atmosphere may slow, in large part due to the increasing temperature difference between the surface and deeper layers of the ocean. The temperature gradient, especially in the open ocean, reduces mixing of carbon with the deep ocean and limits nutrient delivery to the surface, which weakens the BCP but also decreases the return of carbon from depth, thus increasing deep carbon storage.

Much remains to be learned about the magnitude and sign of the expected changes in both mixing and biology, and where in the ocean these changes will be most significant and why. More and better measurements are needed to answer these questions, including seasonally resolved observations, and more accurate and complex models, including improved representation of deep-ocean and seafloor processes, as well as machine learning approaches to fill in gaps where data are missing or sparse.

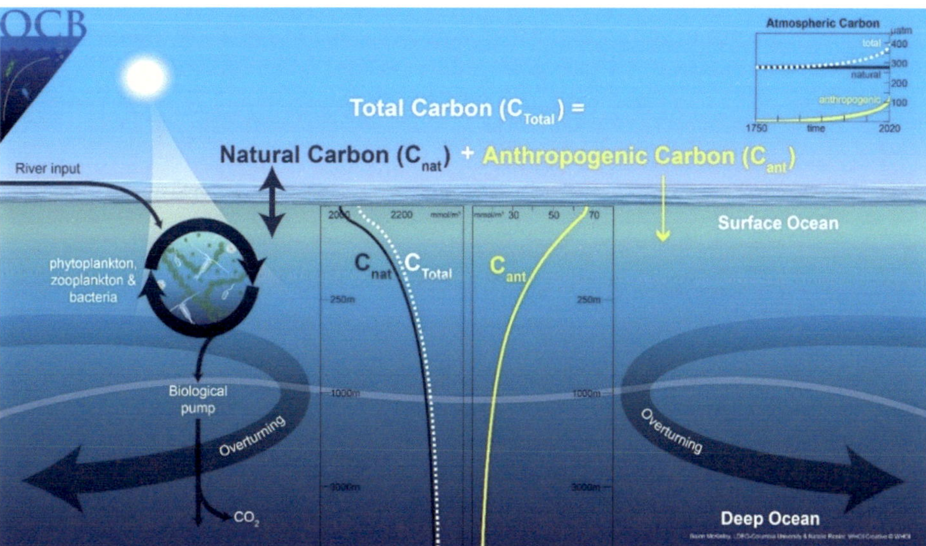

FIGURE 2.3 Ocean carbon cycles.
SOURCE: Galen McKinley, Lamont-Doherty Earth Observatory, Columbia University, and Natalie Renier, Woods Hole Oceanic Institution Creative Studio.

Overall, the Southern Ocean has an outsized impact on the atmospheric CO_2 inventory because it is where the ocean's largest CO_2 reservoir comes into direct and rapid contact with the atmosphere (Gray, 2024). The importance of this region on ocean CO_2 uptake is uncertain (Chikamoto and DeNezio, 2021; Gallego et al., 2020; Zhong and Rojanasakul, 2024); sustained observations in this region will be key to predicting the sensitivity of the ocean carbon sink.

The latest experiments with coupled ocean–atmosphere models indicate that the ocean's ability to absorb carbon will likely decline, but the magnitude of this decline varies between models (Gruber et al., 2023; Wilson et al., 2022). Model uncertainty reflects the fact that the climate system and the carbon cycle are coupled by interconnected processes, making the task of developing accurate forecasts challenging. For example, at the most basic level, as ocean temperature and acidification increase, the capacity of seawater to store carbon decreases. While relatively small positive feedbacks are easy to represent in the models, more substantial changes, which are expected in the next decades, are much harder to represent. Specifically, as the surface ocean warms, thermal stratification slows vertical mixing, both annually and seasonally, particularly at high latitudes. This has multiple impacts, one of which is to weaken the ocean's biological carbon pump and thus the large-scale transfer of carbon in sinking particles from the surface ocean to the deep sea (see Box 2.2; Wilson et al., 2022). Changes in the biological pump and ocean mixing could also exacerbate the release of carbon into the atmosphere at some locations (Gruber et al., 2023), and the ongoing acidification of the ocean (Carter et al., 2017; Gregor and Gruber, 2021) will reduce the ocean's chemical ability to absorb carbon from the atmosphere (Figure 2.4). Addressing these and other factors will be critical for reducing the uncertainty in forecasting trends in atmospheric CO_2 under various emission scenarios, as well as for identifying potential tipping points where the rate of ocean carbon uptake changes dramatically.

Understanding the Ocean Heat Sink

The representation of ocean heat uptake and distribution in coupled ocean–atmosphere models is still a work in progress. Globally, GO-SHIP repeated ship measurements (occupying stations every 5–10 years), and the core Argo array have provided the most definitive data on the ocean's uptake of excess warming (von Schuckmann et al., 2023), in part because these measurements provide an unprecedented look at what is happening in the ocean's interior (Figure 2.5) in space and in time. The core Argo array provides the highest frequency of observations but is primarily restricted to the upper 2000 meters. Because the oceanographic community has recognized the importance of sustained observations over the full ocean depths, Deep Argo[2] was developed by Scripps Institution of Oceanography and Teledyne-Webb Research. Such measurements may be important for reconciling the discrepancy between ocean heat uptake measured by in situ methods and those estimated from satellite observations of ocean thermal expansion (Marti et al., 2024).

Based on spatially detailed observations of ocean heat content as measured by Argo autonomous floats beginning in 2005, a picture is emerging of how the accumulated heat is being distributed across each basin from the surface to intermediate depths (700–2000 m; Figure 2.5; Li et al., 2023; von Schuckmann et al., 2023). Such detailed observational constraints have been critical for testing and improving dynamical models and thus identifying the need for improved representation of key physical processes that regulate heat fluxes on various scales.

Research Advancements Needed to Understand the Ocean's Ability to Absorb Heat and Carbon

Role of Global Circulation

At the largest scale, freshwater and heat are transported by large-scale ocean circulation (e.g., AMOC). However, some models suggest that a slowdown in AMOC will decrease the efficiency of this heat transport

---

[2] See https://argo.ucsd.edu/expansion/deep-argo-mission/ (accessed January 22, 2025).

(Curtis and Fedorov, 2024; Mecking and Drijfhout, 2023; van Westen et al., 2024). Based on model predictions, the most recent (sixth) annual report from the Intergovernmental Panel on Climate Change (IPCC, 2023) indicates there is a high likelihood that AMOC will decline sometime in the 21st century. The report also states that "there is medium confidence that the Atlantic Meridional Overturning Circulation will not collapse abruptly before 2100" (IPCC, 2023, p. 18). However, questions remain as to how much confidence to place in the models' forecasting abilities, given their relatively poor ability to simulate past AMOC activity and the lack of observational evidence of AMOC decline (McCarthy and Caesar, 2023). Observations of the past recently collected from sediment cores have shown that the sea ice extent in the Southern Ocean can drive high-amplitude, millennial-scale variations in the strength of the Antarctic Circumpolar Current, which can also influence AMOC by modulating the Pacific–Atlantic exchange of ocean water (Wu et al., 2021). These studies have demonstrated the importance of understanding change at both poles and in the Atlantic and Pacific oceans in order to better forecast the tipping point where AMOC reaches a new stable state, perhaps at a decreased strength from what is observed today.

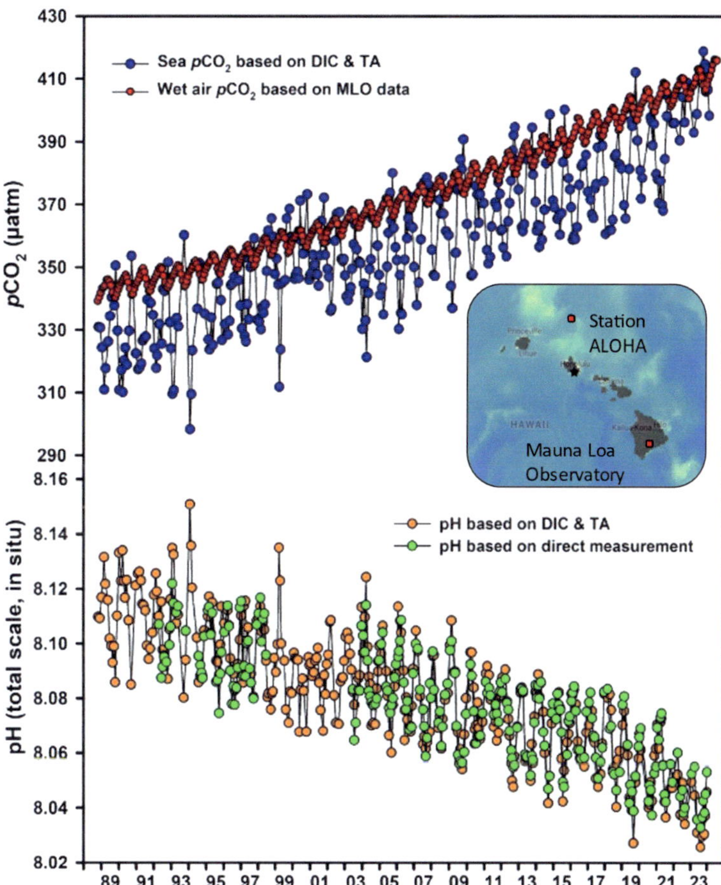

FIGURE 2.4 Carbon dioxide and pH changes at Station ALOHA in the North Pacific, from 1958 to 2024.
NOTES: DIC = dissolved inorganic carbon; MLO = Mauna Loa Observatory; pCO_2 = partial pressure of carbon dioxide; TA = total alkalinity.
SOURCE: Angelicque White. Data: Measurements by the Scripps CO_2 Program and the Schmidt Ocean Institute are supported by the U.S. National Science Foundation (NSF); by the Eric and Wendy Schmidt Fund for Strategic Innovation; and by Earth Networks, a technology company collaborating with Scripps to expand the global greenhouse gas monitoring network. In-kind support for field operations is also provided by the National Oceanographic and Atmospheric Administration, NSF, Environment and Climate Change Canada, and the New Zealand National Institute for Water and Atmospheric Research.

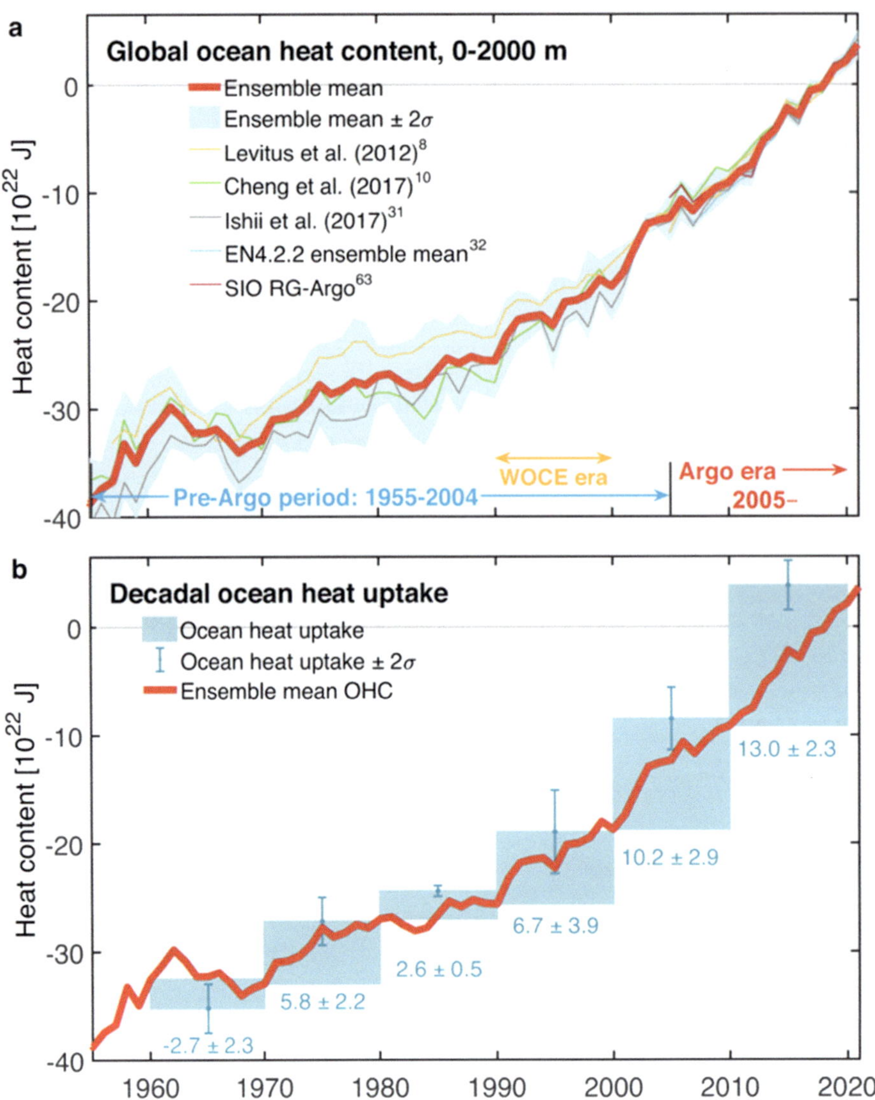

FIGURE 2.5 Multidecadal acceleration in global ocean warming. NOTES: Ocean heat content over time relative to the 2016–2020 mean. The red lines in both panels represent the ensemble mean time series of global ocean heat content (OHC), 1955–2020. SIO RG Argo = Scripps Institution of Oceanography Roemmich-Gilson Argo Climatology; WOCE = World Ocean Circulation Experiment.
SOURCE: Li et al., 2023.

Ice Sheet–Ocean Interactions

Understanding the stability of the Antarctic and Greenland ice sheets is crucial for predicting future sea level rise and deep-ocean circulation. For the Southern Ocean in particular, input of meltwater can alter local circulation, which in turn can impact the role of the Southern Ocean in heat and carbon exchange with the atmosphere. For example, glacial meltwater input can increase stratification and prevent the release of heat from the Southern Ocean interior, which warms subsurface waters further. Warming subsurface waters can act as a positive feedback, further increasing ice sheet melting (Bronselaer et al., 2018). For marine-terminating glaciers, the *grounding zone*—where ice sheets transition from being grounded on the continent to a floating ice shelf—is particularly susceptible to ocean warming, which can exacerbate mass loss and lead to further meltwater input (Bradley and Hewitt, 2024; Payne et al., 2021; Robel et al., 2022), So far, the role of warmwater intrusions near grounding zones, and whether future inputs of meltwater will strengthen or weaken subsurface warming, has been primarily explored in models (e.g., Beadling et al., 2022; Bradley and Hewitt, 2024). Yet the model-predicted sensitivity of ice sheets to warmwater intrusions (e.g., Bradley and Hewitt, 2024) needs to be better constrained with observations around and upstream of grounding zones, and the role of meltwater in controlling the exchange of water between the Antarctic shelf and open ocean requires a better characterization of the currents and cross-slope exchange in this region (Beadling et al., 2022). Water column structure in and around the Southern Ocean impacts the physical processes that drive heat and carbon exchange but can also alter the efficiency and major pathways of the biological carbon pump (Lacour et al., 2023), with consequences that are currently not well understood but may be predicted from past records of Southern Ocean productivity.

Ocean Surface Turbulence

On the other end of the scale of physical processes that impact ocean heat uptake are small-scale turbulent processes near the ocean surface that act to control the transfer of heat, moisture, and dissolved greenhouse gases between the ocean and atmosphere (Dong et al., 2024; MacKinnon et al., 2016, 2021; Seo et al., 2023; Su et al., 2018; Whalen et al., 2020; Wu and Mahdevan, 2024; Yu, 2019). Although global models generally parameterize these as one-dimensional mixing processes (on both the ocean and atmospheric side of the interface), recent observations have shown that turbulence often has a fully three-dimensional geography, varies substantially on meso- and submeso-spatial scales and can lead to strong nonlinear coupling between the ocean and atmosphere on scales that are poorly measured and understood. Resolving these gaps will require new types of intensive, coordinated, multiplatform observations and new numerical and analysis techniques (including machine learning) to be depicted in the models.

These and other advances will help reduce the uncertainties in model forecasts of how carbon and heat uptake may change in the future, globally and spatially. Such an enhanced understanding will also contribute to improving model forecasts of changes in ocean circulation and other processes that affect marine ecosystems. As such, understanding potential changes in carbon and heat uptake represents both a vital and an urgent priority (as defined in Box 1.3 in Chapter 1) for the ocean science research community.

Societal Relevance and Broader Impacts of This Research

It cannot be assumed that the ocean will continue to provide the key climate, heat, carbon and associated societal services that it does presently. There is clear evidence that changes in the efficiency of these processes are very likely to occur and that, while the extent and character of these changes are uncertain, significant disruptions to ocean heat transport and carbon uptake are inevitable. It will be necessary for society to evaluate such changes in the ocean system as it considers the impacts of Earth system changes that are likely in the coming decades.

Marine Carbon Dioxide Removal

The growing field of ocean geoengineering aims to increase the amount of atmospheric carbon uptake by the ocean by enhancing the ocean's natural capacity to absorb and sequester carbon, referred to as marine carbon dioxide removal (mCDR; NASEM, 2022a). Many groups are already exploring mCDR techniques but from a limited knowledge base. There is little clarity on how to determine the success of such experiments, the associated hazards and unintended consequences, or the wider societal benefit that may or may not be associated with their widespread application (NASEM, 2022a). The ocean science research community has a critical role to play in conducting basic research to answer fundamental questions regarding the viability of any mCDR approach, including continuing long-term observations on the ocean's natural carbon uptake mechanisms and their impact on ecosystems, as well as developing tools to verify the efficiency and effects of proposed mCDR strategies. Important questions for researchers to address are whether the ocean's natural CO_2 removal and sequestration processes can be sped up safely and at a scale that would help reduce the emissions deficit; whether the scale needed to make an impact is possible; and, if scale-up is possible, how the carbon cycle processes can be measured to verify removal of carbon dioxide from the atmosphere. Transdisciplinary research through strategic partnerships is needed to evaluate the cost-benefit trade-offs for the ecosystem and for the communities affected by potential mCDR activity.

Earth System Science

The research proposed within the ocean and climate theme will stimulate the development of new understanding of physical, biological, geological, and chemical processes in the ocean and the links between these fundamental components of the Earth system. Other connected disciplines will include engineering (new platforms and technology), mathematics, computational and computer science (data analysis and modeling), biology (organismal issues), and many others. Additionally, the cross-cutting issues that will need to be addressed to answer the research questions in this theme will require new, transdisciplinary studies that will be supported by the training and education of a new cadre of early career researchers. The results of this work will support the growing understanding and awareness of society as to the scope and scale of the changing ocean and what this means in terms of weather patterns and extremes, sea level rise, and the health of coastal waters and the nutrients harvested within. The research will lead to understanding that may guide the development of new policy around mitigation and adaptation strategies for responding to those changes.

Observations, Tools, Technologies, and Human Resource Needs

The observational needs of this research endeavor will require developing new sensors to measure key ocean variables and applying these sensors on platforms, such as Argo floats and other autonomous observing systems, that are already in use, as well as new platforms and approaches to remote ocean observing. These innovative developments will stimulate the commercial sector to further market new technology for the growing need to measure variables that can quantify heat and carbon uptake and redistribution in the global ocean.

Regional- and Global-Scale Observations

A wide range of process studies in key regions needs to be combined with regional- and global-scale surveys for measuring key ocean variables, including heat, salinity, carbon, oxygen, and other physical, biogeochemical, and ecosystem variables. These measurements will enable better understanding of the ocean, which will support the calculation of key fluxes used in regional- and global-scale forecast models. The data from these large-scale surveys can be used to both initialize and calibrate these models. A good example of such a global-scale survey is the autonomous float program Argo, which is presently being expanded from a primary focus on measuring only the surface heat and salinity (Argo) to include biogeochemical measurements (BGC-Argo), acoustics (Passive Acoustic Listeners), and measurements over the full ocean depth (Deep Argo). However, full process studies, which are necessary for quantifying key ecosystem fluxes, will still need to rely on the use of a modernized U.S. Academic Research Fleet for more specialized measurements.

Model Development and Data Management

A key part of this work will be developing new models that both build upon existing models and apply new modeling techniques, including innovation associated with using high-performance computing software and artificial intelligence. Investments in collecting measurements on different temporal and spatial scales—including global circulation and air–sea exchange—and data curation to support new advances in data assimilation, machine learning, and hybrid methods are particularly relevant to advancing these models. The models needed to do this work include high-resolution regional models paired with biogeochemical and biological models and coupled ocean–atmosphere climate models that also include key biological and biogeochemical systems. No single-model architecture will be able to ingest and analyze the amount of data needed to understand these questions. This enhanced modeling will require better coordination, integration, and curation of data from all sources, as well as collaboration with other agencies and international partners for an effective approach to data management.

Paleoceanographic Records

Long-term datasets are necessary for deciphering the periodicity of natural climate variability that occurs on timescales that can obscure the changes associated with anthropogenic warming. Furthermore, the climate periods that are most analogous to current global temperatures and/or atmospheric CO_2 concentrations are more than 100,000 years in the past and as far back as several million years in the past (Anderson et al., 2024; Shackleton et al., 2020). In this context, direct observations of global climate from less than a century ago provide too little data to adequately assess whether the advanced models are accurately simulating Earth's climate at greenhouse gas levels significantly higher (or lower) than present. Most notably, climate archives preserved in subseafloor sediment cores recovered by scientific ocean drilling are critical to ground-truthing theories on the dynamics/coupling of the climate system and the carbon cycle (see Figure 3.4 in the Interim Report [NASEM, 2024b]), as well as the nature of tipping points. For example, evidence continues to emerge on key details (timing/patterns and lead–lag relationships) of past changes in ocean overturning circulation, particularly the history of AMOC, a major contributor to heat redistribution over the last several million years of glacial–interglacial cycles. This information, along with new paleoceanographic data on changes in North Atlantic and Arctic sea ice distribution, are enabling identification of the temperature and salinity thresholds that trigger major shifts in the intensity of AMOC (e.g., Ait Brahim et al., 2022; Walczak et al., 2020; Wu et al., 2021). These findings, along with new ocean drilling records from geographically sensitive regions and analogous time periods and more detailed modern observations, will be key to testing and improving models that are forecasting a rapid reduction in the strength of AMOC in the coming decades and for forecasting impacts on regional climates (e.g., cooling of England and Scandinavia and possible hydrological shifts, such as monsoon system changes).

Additional paleoceanographic evidence of the modes of ocean circulation present in a warmer world, as well as the ocean's role in regulating the carbon cycle (and climate), can come from scientific ocean drilling investigations of more ancient episodes (e.g., early Cenozoic and Cretaceous) of extreme greenhouse periods in Earth's past. For example, sediment cores recovered from the Antarctic margin and elsewhere are providing critical constraints on the evolution of the East and West Antarctic ice sheets (e.g., Marschalek et al., 2021), which—along with the latest reconstructions of past variations in greenhouse gas levels (The Cenozoic CO_2 Proxy Integration Project Consortium, 2023)—point to the potential for rapid responses (i.e., tipping points) to future warming. Further constraining the rates of change and the processes that trigger rapid ice sheet decay, such as interactions between warm(er) ocean currents (resulting from circulation changes) and marginal marine ice sheets and ice shelves (Sproson et al., 2022), are important areas for ongoing ocean science research.

Novel Sensors and Platforms

As discussed above, process and regional studies of heat and carbon will require the development of new sensor and platform systems to support the observational programs to collect the data needed. For example, biogeochemical sensors used to measure carbon are either very limited or unavailable. While some approaches, such as the Wendy Schmidt Ocean Health XPRIZE for pH measurement (Okazaki et al., 2017), have been effective in developing a particular sensor type (making a specific type of measurement), such innovation often takes years to decades and has not always led to quick commercial adoption followed by widespread availability of the critically needed sensors. Those making efforts to develop new technology need to look beyond the initial development phase and consider making such systems available to the ocean observers seeking to make the measurements.

Geographic Priorities

Regional studies need to focus on specific areas relevant to ocean climate and carbon cycling, including the convection regions of the North Atlantic; the Southern Ocean; upwelling regions (e.g., Eastern Boundary Currents and the equatorial Pacific Ocean); equatorial regions (e.g., hurricane formation sites in the tropical Atlantic); and coastal and estuarine regions, where freshwater mixing stimulates production. Those designing such region-focused studies need to carefully consider the balance between having a large-scale geographic focus on inventory analysis (e.g., surveys of carbon and nutrient concentrations) and ground-truthing with the need for process studies (e.g., quantifying carbon fluxes to the deep ocean).

Coupled Physical–Biological Models

In addition, carbon budgets are intrinsically linked to ocean life: the dynamics, ranges, and abundance of species, from viruses to whales. Models that include the species that affect carbon cycling, acquisition, and release are fundamental to success, as are models that parameterize fluxes within food webs. There is a need for further model development on ways to assess, measure, and sense ocean life at the species level in the world's ocean and incorporate this understanding into the models in such a way that they are accurately represented as parameters in the climate models.

Workforce Development

Common to all three urgent research themes, and discussed further in Chapter 3, the groundbreaking research required to resolve these societal challenges will require new cohorts of transdisciplinary researchers and ocean management professionals. In particular, local communities and their resource managers play a crucial role in directing effective research and in using research to formulate practical solutions to local problems. New approaches to providing education, training, and development are key to the scientific process, including in regional and global scale studies. The transdisciplinarity of these themes spans the classical branches of ocean science—biological, chemical, geological, and physical—but also includes ocean engineering, genomics, and computer science, as well as—given the themes' societal relevance—social sciences and humanities, when appropriate. The need for transdisciplinary research to be collaboratively developed with multiple interest holders and communities is described further in Chapter 3.

ECOSYSTEM RESILIENCE
How will marine ecosystems respond to changes in the Earth system?

Ecosystem resilience, broadly speaking, refers to the capacity of an ecosystem to absorb repeated disturbances or shocks and adapt to change without fundamentally switching to an alternative stable state (Holling, 1973). One of the most highly referenced research papers on the topic of coastal ecosystems discussed the existential threats that disturbances—in particular, eutrophication and overfishing—pose to marine food webs (Jackson et al., 2001). This and other research studies emphasize the connectivity among organisms and environment, and how environmental impacts can lead to ecosystem state change, including collapse. Drawing on evidence from highly studied ecosystems and long-term datasets, these studies increased understanding of how ecosystems resist and recover from environmental change, and the points at which they can shift to a highly altered state (Heinze et al., 2021). The message of this and related research papers is that life in the ocean will go on despite tragic impacts, but it may do so in a manner so different from present conditions that it lacks much of what humanity values.

Understanding the resilience of marine ecosystems to environmental changes remains a significant challenge. A key component of the challenge of understanding ecosystem resilience is deciphering the patterns of species' distribution and abundance across space and time, commonly referred to as biogeography, which depend on species-specific interactions, growth rates, predation, and environment changes and resource availability. Early advances in marine ecology have provided foundational concepts of how ecosystems function, including the ideas that the presence of some "keystone" organisms play an outsized role in the function of whole systems, that species diversity increases ecosystem resilience, and that microbes play a fundamental role. These fundamental theories benefit from examination in the context of the wide range of habitats and rapidly changing conditions in the global ocean.

It is also important to consider the physical, chemical, and biological connectivity among marine ecosystems. There are many examples of the depth of interconnection among variable ecosystems and how they bolster one another. For one, vertically migrating animals mix stratified waters (e.g., the epi- and mesopelagic) and move nutrients and carbon between deep and surface zones, providing nutrition for bathypelagic and benthic organisms (Wirtz and Smith, 2020). Coastal coral reefs, mangrove forests, and seagrass beds and marshes support seashore communities through nursery protection of young fish, nutrient provision, but also by dissipating wave energy delivered to shorelines and minimizing damage. Equally important viruses, bacteria and archaea, and protists play critical roles in the ocean biogeochemistry that affects nutrient availability and productivity among all ecosystems (e.g., Holt et al., 2023; Shah Walter et al., 2018; Suttle, 2007).

These complex issues of ecosystem stability play a huge role in impacts of marine fisheries, coastal habitats, biodiversity, and global climate impacts on humanity. Advancing the understanding of those factors that confer or undercut ecosystem resilience also enhances humanity's ability to manage—and when necessary, adapt to—changes in marine ecosystem functionality.

The challenge for the next decade is to understand biological communities and their functions in the ocean over multiple temporal and spatial scales. These understandings serve as both sentinels to change and ciphers for how change will emerge under different environmental pressures. This includes the need to predict when ecosystems can and cannot recover from disturbances. Not only is it essential to understand how ecosystems have responded in the past, but there is a strong need to develop the ability to forecast how ecosystem structure function and biodiversity will respond to changes in the Earth system in the near future. As a result, scientific research to establish a basis for change trajectories in ecosystems is vital, and research to forecast where trajectories lead for these ecosystems, especially those that humans depend upon, is urgent. Addressing these themes through a transdisciplinary research lens will create the foundation for understanding of how changing physical, climate, and social environments affect the interdependent relationships between humans and ocean.

Background and Context

What enables a marine ecosystem to be resilient in response to change is an area of significant scientific inquiry (e.g., Ma et al., 2021). Foundational questions remain, in both well-studied systems such as temperate fisheries, coral reefs, and marshes, and less-studied areas such as the deep ocean and hard-to-reach locations (e.g., beneath sea ice in polar regions, seamounts). Ongoing NSF investments in ocean monitoring and long-term ecological studies help set the context for understanding how species and ecosystems respond to natural and anthropogenic environmental change.

Key Concepts in Understanding Ecosystem Resilience

In conducting research on resilience, the three fundamental "Rs"—resilience, redundancy, and resistance—are key to understanding how change will impact ecosystems and the services they provide (Levin and Lubchenco, 2008). Ecological *resilience* focuses on the capacity of an ecosystem to absorb and adapt to disturbances while maintaining typical structure and function (components and processes). Closely affiliated with resilience is *redundancy,* the ability of one taxon to play the same important ecosystem role as another, and *resistance,* the ability for taxonomic composition to remain unchanged in the face of a disturbance (Allison and Martiny, 2008; Levin and Lubchenco, 2008). The facets that delineate variability in these three Rs also warrant more attention.

A parallel concept to resilience is *tipping points* (Selkoe et al., 2015), or the stress levels at which ecosystems shift rapidly in terms of biological structure and function. Using these scientific concepts to learn more about the dynamics of ocean ecosystems will translate directly to reducing the uncertainty in predictive models forecasting when and how ecosystems change. Accordingly, this research topic is a priority for the ocean science community, spread across myriad ecosystems and millions of species that make up ocean life.

The Importance of Biodiversity in Conferring Ecosystem Resilience

One important facet underpinning ecosystem resilience is the biodiversity inherent within an ecosystem. Biodiversity effects on ecosystem resilience have been a pivotal topic over the past decade, with a series of studies showing that resilience to environmental change is bolstered in habitats with a large number of species (Duffy et al., 2016; Gamfeldt et al., 2015; O'Leary et al., 2017; Palumbi et al., 2009; Stachowicz et al., 2007; Vasileiadou et al., 2024; Worm et al., 2006). Moreover, the rate at which ecosystems recover from perturbations may be accelerated in habitats with greater biodiversity (Lotze et al., 2011). Specifically, it has been posited that ecosystem resilience is conferred when a diversity of organisms in a given habitat can carry out the same ecosystem processes or roles . This functional redundancy has been studied across many habitats using food web analysis, lipidomics, metatransciptomics, and metagenomics (e.g., Allison and Martiny, 2008; Galand et al., 2018; McParland et al., 2024; Sher et al., 2024). Continued studies of functionality are necessary, however, to identify ecological tipping points, where ecosystems can no longer recover to a prior state because of structural shifts in community composition that result from a loss of both biodiversity and function. Such research may be especially important in coastal habitats such as coral reefs, salt marshes, and mangroves where an inability to recover has a major effect on human coastal communities (e.g., Laurance, 2013). More remote ecosystems, such as deep-sea benthic systems, are also impacted by human activity connected to natural resource extraction (oil and gas, critical minerals) and may respond in ways that are difficult to predict, given the relatively more limited study of these places. The *Deepwater Horizon* oil spill is an excellent example of the importance of addressing knowledge gaps in ecosystem resilience in the deep sea to enhance science-based decision making.

While the understanding of these processes in animal communities is growing, the role of biodiversity and functional redundancy in marine microbial communities is not as well understood for a variety of reasons. First, the physiological repertoire of microbes (specifically, bacteria and archaea) is far greater than that of animals. Second, the mobility of genes among microbes may provide greater resilience to, for example, viral infection within a microbial mat (Hwang et al., 2023). In some cases, microbial communities maintain high biodiversity even when carrying out a limited set of metabolic pathways, which may confer resilience (Louca et al., 2018). How resource-limited conditions—such as the open, oligotrophic ocean—can maintain unexpectedly high diversity is paradoxical (Galand et al., 2018), highlighting the importance of further studies of resilience across taxonomic domains.

The important role of biodiversity in the ocean contrasts starkly with the increasing threats to biodiversity and unprecedented rates of decline. Worldwide, more than 25 percent of plant and animal species are estimated to be at risk of extinction (International Union for Conservation of Nature [IUCN], 2012). In the ocean, unsustainable fishing practices, climate change, and pollution are leading drivers of biodiversity loss within ecosystems (Jaureguiberry et al., 2022). These losses are characterized by declines in population sizes of many species and extents, as well as the growth of "ecological generalists and disturbance-adapted species" in lieu of endemic species (IPBES, 2019, p. 3). In light of the aforementioned research relating biodiversity to ecosystem resilience, such losses raise the possibility that ecosystems will continue to be more vulnerable as rates of extinction increase. Improving understanding of direct and indirect drivers of biodiversity loss, as well as means of its prevention and remediation, is a key scientific priority in ocean science.

Molecular Genetics Tools for Understanding Ecosystem Resilience

Global surveys—such as the yearlong expeditions of the research schooner *Tara*—demonstrate how ocean biodiversity can be assayed using combinations of traditional tools (e.g., morphological) and evolving molecular genetics tools (Sunagawa et al., 2020). In the past decade, however, the use of molecular tools to study ecosystems has expanded through use of DNA for monitoring (Garland et al., 2023; Shum et al., 2019; Stat et al., 2017), strengthening the ability to enumerate biodiversity across taxonomic groups, and thus providing quantitative, reproducible data on the relationship between biodiversity and ecosystem resilience. Such tools have also been used to measure ancient eukaryotic DNA in ocean sediments (*seda*DNA). These new measurements open the possibility to understand ecosystem-wide changes and paleoenvironmental insights across major climatic transitions, including past eras of rapid ocean warming (Armbrecht et al., 2022; Crump, 2021; Harðardóttir et al., 2024). These recent results also establish a relationship between habitat biodiversity and the resilience and productivity of marine ecosystems.

Biodiversity, Resilience, and Connectivity in Distinct Ocean Habitats

Deep Sea Ocean depths below 200 meters (meso, bathy, and abysso pelagic zones) represent Earth's largest habitat and harbor tremendous animal and microbial biodiversity (Paulus, 2021; Rex and Etter, 2010). Some studies suggest that only two-thirds of deep-sea biodiversity has been identified, yet these and remaining undiscovered species are under threat from anthropogenic activities (Costello and Chaudhary, 2017). Moreover, the deep sea provides a stunning array of ecosystem services that support upper-ocean productivity, including commercially relevant activities (Jobstvogt et al., 2014; Orcutt et al., 2020; Thurber et al., 2014).

Coral Reefs Coral reefs constitute only about 1 percent of the ocean surface but are home to about 25 percent of known ocean species. They tend to have much higher productivity than expected for oligotrophic waters, largely because of the symbiosis between reef-building corals and dinoflagellates (Symbiodiniaceae). Scleractinian corals have been a dominant ecological taxon for 250 million years, yet they have been strongly affected by past planetary periods of high heat and low pH (e.g., the Paleocene-Eocene Thermal Maximum; Kiessling et al., 2024), making them an important group to study climate impacts on biodiversity and resilience. Reefs also are affected strongly by current climate driven heat waves and are a model system for understanding the potential for adaptive evolution (Bay et al., 2017; McManus et al., 2021).

Open Oceans Offshore habitats from the poles to the tropics shelter a wide variety of species fundamental to fisheries, carbon flow, and production by photosynthetic plankton. The microbial loop in the open ocean can play a large role in carbon cycling within and between systems (see Gilbert and Mitra, 2022). It is also a place where the transient movement of organisms, including large migratory whales, sharks, fish, and turtles, impacts system resilience and interactions.

Polar Seas Both Arctic and Antarctic habitats have a wealth of species adapted to strong seasonal changes in productivity and to cold conditions. These span microbiomes, microplankton, intermediate grazers such as krill, and migratory predators such as baleen whales. These locations are important for study of ecosystem resilience with examples of recovery, such as whale populations that recovered after aggressive whaling in the 18th, 19th, and 20th centuries (Savoca et al., 2024); such examples can inform future efforts to understand ecosystem resilience.

Resilience and Connectivity Among Marine Ecosystems

Ecosystems resilience is intertwined with the ability to resist change. Such resistance is often connected to the import and export of taxonomic groups and their associated function. For example, sharks connect reef ecosystems (Dixon and Gallagher, 2023), but their overall contribution to ecosystem resilience remains unclear in many cases. Change in the ocean is connected to both direct and indirect modifications to ecosystems (Watson et al., 2018). Direct modifications include the progressive addition of built and modified habitats in the ocean, known as ocean sprawl (Bishop et al., 2017). These additions stem from natural resource (e.g., energy, aquaculture) activities that have grown logarithmically in shelf, slope, and deep-sea environments since the 1980s (Bugnot et al., 2021; McCauley et al., 2015) and are expected to continue to rise, impacting ecological connectivity in populations spanning whales, fish (Firth et al., 2016), and microorganisms (Hampel et al., 2023). Habitat type and density can speed or impede the movement of organisms between habitats (Bishop et al., 2017) and change their biological character. Macroscale installations on the coasts and seabed (e.g., coastal and energy infrastructure, shipwrecks) are also joined by debris accumulating in the ocean for thousands of years (Amon et al., 2020; Lebreton et al., 2024) with unknown ecosystem impacts, and that may confound focused studies of natural history of naturally occurring deep-sea habitats (e.g., seeps, vents, seamounts).

An example of an indirect modification is eutrophication and the frequently associated changes in the supply and form of organic carbon to the coastal ocean. Such changes have been shown to impact taxonomic composition and life habitat (free living vs. particle attached) of bacterioplankton (Crump et al., 1999), which may be consequential to turnover of carbon in the ocean through the microbial loop, the expansion of oxygen minimum zones, and the export of organic carbon to the deep ocean. How eutrophication interacts with the movement of organisms and how it influences ecosystem resilience are open questions, even after decades of study, that rise in importance when microhabitats, including plastics, are considered.

Societal Relevance and Broader Impacts of the Research

The ocean maintains the habitability of Earth's biosphere and confers environmental stability upon which humanity depends heavily. The next decade of studying marine ecosystems must include forecasting how changing marine environmental conditions will affect ecosystem resilience and thus impact human communities (and vice versa).

Basic research that informs topics affecting the blue economy, such as food security and long-term resource sustainability, may be particularly valuable. For example, targeted short- and long-term studies of ocean warming, acidification, and deoxygenation across a variety of spatial and temporal scales can help provide the information necessary to anticipate adverse effect on marine ecosystems and the downstream consequences of those effects on other communities, including humankind (see Feeley et al., 2010). Moreover, developing more robust models for predicting the weakening of the biological pump/export production—and how this will affect trophic structure, function, and movement of organisms—is key. Such models require fine-grained understanding of the full suites of ocean biodiversity and how these complex communities intersect with environmental conditions. While this may be aspirational, it prioritizes research areas and will help in conservation efforts, coastal protection, fisheries yields, and food security (e.g., Laurenceau-Cornec et al., 2023). Co-developing technologies for rapid measurement and assessment of ocean biological and functional diversity could be prioritized, specifically technologies that allow for more rapid, higher-fidelity assessments of species-level biodiversity (e.g., advanced environmental DNA technologies [Takahashi et al., 2023]), as well as function (e.g., advanced technologies for measuring metabolic processes). Understanding matter and energy transfer in and across marine ecosystems, as well as to the continents, will help us better identify the environmental stressors that threaten resilience (e.g., Hoehler et al.,

2023). Finally, studies that better quantify the coupling of benthic and pelagic processes, as well as the coupling across pelagic zones, are essential to understanding how marine ecosystems are linked across ocean biomes. Moreover, they will provide the information necessary to understand the nature and extent of anthropogenic activities—such as deep-sea mining and its midwater pollution—on scales that transcend single-ocean ecosystems (Box 2.3; Drazen et al., 2020). NSF is uniquely situated to support basic research, including that with direct relevance to application, to establish a framework for thoughtful, well-informed discussions about anthropogenic impacts on marine ecosystems.

BOX 2.3
Potential Impacts of Deep-Sea Mining

FIGURE 2.6 A coral garden observed at a depth of 2,365 meters.
SOURCE: National Oceanic and Atmospheric Administration Office of Exploration and Research, Deep-Sea Symphony: Exploring the Musicians Seamounts.

Deep-sea mining likely has an outsized impact on deep-sea animals, which are adapted to the low-nutrient, low-disturbance conditions (Figure 2.6). These organisms are not equipped with the physiological or biochemical capacity to cope with increased suspended particles, changes in pH or oxygen, and physical perturbation that will almost certainly arise from seabed mining (e.g., Grégoire et al., 2023). Moreover, mining will very likely produce particulate plumes that will reside in the overlying water column (Christiansen et al., 2020). These nano- and microparticles of minerals can have severe impacts on pelagic organisms, especially larva (Drazen et al., 2020). It is equally likely that suspended mining particles will impact commercially relevant species as well. A number of small-scale seabed mining studies support these conclusions (Voonahme et al., 2020), but broadly speaking, there is a massive lack of baseline data (Le et al., 2022), making it difficult to provide quantitative estimates of these impacts and making it easier for mining advocates to misrepresent or disregard existing data.

The demand for rare earth elements will likely fuel continued interest in deep-sea mining. Developing a robust, comprehensive understanding of the potential benefits and costs of seabed mining depends on continued deep-sea research, including on the connectivity between benthic and pelagic zones. Studies to date suggest that the impacts of seabed mining have the potential to be widespread, including impacts on deep-sea biodiversity, deep-sea ecosystem services, and comparable impacts on mid- and upper-ocean ecosystems.

Techniques, Tools, Technologies, and Human Resource Needs

Defining the key functions of healthy ecosystems and understanding what enables an ecosystem to be resilient requires multiple perspectives from the experimental to the societal and will take advantage of a varying set of skills, interests, and investments in capacity. A broad swath of data from many disciplines is required to convey a cohesive understanding of facets that govern ecosystem health. For example, understanding the ecosystem will require extensive data processing of biodiversity patterns across taxa for the dynamic patterns to seed future forecasts and mechanistic hypothesis testing.

Measuring Biodiversity Resilience

The explosion of molecular tools in the past decade has expanded the capability to monitor whole-community dynamics spanning all three domains on the tree of life: bacteria, archaea, and eukaryotes, and to simultaneously delve taxonomic composition, potential function, and expression of function. Tools that provide results of this nature—such as advancements in environmental and organismal DNA, -omics, and bioinformatics—need to be explored and developed to expand monitoring capacity (Chen et al., 2024; Hendricks et al., 2023). Such capability is necessary if metrics of biodiversity are to be included in forecasting products, and to measure and communicate how ocean change directly affects ecosystem services supported by biodiverse marine ecosystems. However, these techniques remain time and resource intensive and do not provide instantaneous or in situ data. Biosensors for specific taxa (e.g., harmful algal blooms, copepods, pelagic fish, many others) need to be developed to provide dynamic measurements of specific ecosystem components.

Highly productive ecosystems often have lower alpha diversity (e.g., hydrothermal vents and seeps, upwelling ecosystems) due to dominance by few taxa; and research strategies must therefore incorporate concepts of bloom dynamics alongside productivity to better understand biodiversity patterns. Moreover, there is a paucity of technologies for measuring metabolic rates in situ, especially automated technologies. Such technologies would enable a more direct assessment of function and activity that could be related to biodiversity indirectly. In addition, artificial intelligence–enabled video documentation and annotation, coupled with molecular techniques, may further enhance the ability to quickly assess patterns of biodiversity across space and time.

Food Web Dynamics and Emergent Ecosystem Properties

Food web dynamics are fundamental reflections of ecosystem function and resilience. Food web construction and exploration relies on correct identification of species and their diets in complex marine ecosystems (Paine, 1966). This is a serious challenge in complex ecosystems with thousands of predatory and prey species. This problem has required significant simplification of datasets in the past but can now be better managed using the molecular techniques mentioned above. When those approaches are coupled with trait-based or theoretical-based estimates of species interactions (Brose et al., 2019; Krumhardt et al., 2024; McGill et al., 2006; van Denderen et al., 2021) and biogeochemical measurements, they may afford enhanced opportunities to construct marine food webs with greater depth and realism.

Modeling marine food web interactions with improved skill and accuracy is needed in the next decade. In the past, this modeling has relied on simulation testing and sensitivity analysis to advance and critique hypotheses and to calibrate, validate, and verify modeling of marine food web responses to a wide range of phenomena (Coll and Lotze, 2016; Fulton et al., 2011, 2019; Heymans et al., 2014; Libralato et al., 2019). In fact, marine food web models have established international best practices, which are being increasingly considered in applied research (Craig and Link, 2023; Fulton et al., 2011; Heymans et al., 2016). Coupling food web models, however, to other models of oceanographic features remains a challenge, as does coupling them with socioeconomic models. In addition, most food web models do not include microbial components of the ecosystem; if they are included, they are poorly parameterized. Emergent and cybernetic features of systems thinking may also be helpful to understand ecosystem resilience holistically (Sunagawa et al., 2020).

Insights gained from systematic views of marine ecosystems result in perspectives that would have been missed otherwise. Emergent features of marine ecosystems, such as rapid shifts in species composition or abundance, can serve as early warning signals of approaching tipping points (Heinze et al., 2021). Knowing how well and how fast a system will respond to interventions or insults, such as oil spills, is now possible using cybernetic measures of resilience, including ascendancy, network measures, and trophic throughput (Lewis et al., 2021). Cumulative biomass and production inflection points across trophic levels can serve more effectively than species-level evaluation as early warning signals of responses to change (Libralato et al., 2019). Novel works on food webs and emergent properties will provide a more composite and comprehensive evaluation of ecosystem resilience.

Biological Teleconnectivity and the Geographic Scale of Ecosystems

The connectivity of biological populations at both small (Cowen et al., 2007) and large scales also influences ecosystem resilience. For example, animal tracking data show that migrating macrofauna (e.g., whales, seabirds) transit across ocean basins. There is growing evidence that some species' populations can be exchanged and intermingled across these basins, such as the movement of marine invertebrates across the Arctic from the Pacific to the Atlantic during the Last Glacial Maximum (Palumbi and Kessing, 1991). Oceanographic connections through this gateway are projected to expand as the Arctic becomes more ice free, and also in response to decreased seasonal stratification of converging Arctic, Pacific, and Atlantic water masses converging in the Arctic and increased inflow of Atlantic and Pacific waters due to large-scale circulation changes. The ecological implications of these changes to ice cover and stratification are not well known at this time and will be difficult to model and predict without further study.

Many questions remain about connectivity, particularly at the base of the food web, spanning prokaryotic and eukaryotic plankton and how they move across ocean basins. Datasets on genetic diversity can help determine the degree of differentiation in populations across vast distances and the role of currents (Hamdan et al., 2012; White et al., 2010). Currents play a strong role in distribution of adaptive genes and the ability of ocean populations to evolve during rapid environmental change (e.g., McManus et al., 2021). An important step in the next decade is quantifying the global, basin, and regional scales of biological teleconnectivity. This will require developing more advanced ocean flow models and testing their ability to predict biological transport of larvae and juveniles. If global-scale, biological teleconnectivity does exist across the ocean basins, understanding the ramifications of genetic mixing, speciation, microevolutionary changes, food web dynamics, other species interactions, and even human dependencies upon these biota will be needed. The impact of such teleconnections may be further amplified in light of the influence of climate change on species distributions and changing water conditions.

The Role of Marine Paleobiology in Untangling Ecosystem Responses

A challenge to forecasting marine ecosystem responses to changes in the Earth system is that natural multidecadal variabilities in ocean conditions, and short records from direct observations, obscure the detection of longer-term trends (Gulev et al., 2021). Additionally, the best analogs for model scenarios of a near-future warmer world occurred at least 100,000 years in the past to several million years in the past (see Figure 3.4 in NASEM, 2024b; see also Rae et al., 2021). For these reasons, the analysis of high-resolution marine microfossils, molecular fossils (i.e., biomarkers), and *seda*DNA preserved in subseafloor sediment cores collected by ocean drilling are key to understanding patterns of species' distribution and abundance across space and time, changes in productivity, and other ecosystem functions and dynamics. Microfossil datasets provide the means to examine how ecosystems responded in the distant, but highly relevant, past warming ocean and to examine the acidification and deoxygenation of ocean waters. Microfossil records have informed predictive models on the vertical compression of habitats as the ocean warms and their expansion as the ocean cools (e.g., Crichton et al., 2023). Molecular fossils from Cretaceous subseafloor sediments have shown that abrupt, sustained increases in primary production can trigger marine microbial reorganization from the surface waters to the seafloor, destabilizing carbon cycling and promoting progressive marine deoxygenation and ultimately ocean anoxia events (Connock et al., 2022). These paleobiological examples show how changes in plankton ecology and diversity, as well as microbial metabolic activity, could alter the efficiency of the ocean's biological carbon pump (and thus its ability to absorb carbon) (Crichton et al., 2023; Henson et al., 2022). Other ecological concerns—such as determining whether there is an upper thermal limit for tropical habitability of phyto- and zooplankton as the planet warms and the extent to which poleward shifts may represent precursor signals that lead to extinction—would require new marine palaeobiological records to fully understand.

In summary, the next decade will continue to bring change (natural and anthropogenic) that significantly impacts the health and survival of marine organisms, marine ecosystems, and coastal communities. Transdisciplinary science that seeks to address and understand this change is needed, along with innovative and wide-ranging perspectives applied to the research. As a leader in the field of ocean sciences, the United States has a responsibility to facilitate these unique research approaches in this critical time of change.

EXTREME EVENTS
How can the ability to forecast extreme events driven by ocean and seafloor processes be improved?

The ocean provides many benefits to society, but it also generates extreme events that can be devastating to people and ecosystems. Examples range from coastal hazards, such as subduction zone earthquakes, tsunamis, hurricanes, storm surges, and sea level rise, to phenomena that reach inland, such as monsoons, atmospheric rivers, and heat waves; these extreme events are often defining moments for communities. Likewise, extreme biological events, such as red tides, mass fish die-offs, coral bleaching, and the disappearance of 11 billion snow crabs from Alaska waters, can also disrupt or destroy local economies and ecosystems (Kruse, 2023).

While coastal areas, which are home to 40 percent of the U.S. population (NOAA, 2025a), often bear the brunt of ocean-based extreme events, impacts from large atmospheric phenomena, such as El Niño or the North Atlantic Oscillation, can also lead to extreme weather events across the entire continental United States. Developing the ability to forecast such events relies upon foundational scientific knowledge of the underlying causes of extreme events, which in turn will allow researchers to better assess and inform how changes in ocean conditions and ecosystems will impact society in the future. Many extreme events are increasing in frequency and/or intensity, making the need to understand their mechanisms and improve their forecasts both urgent and vital.

Background and Context

Societal vulnerability to extreme events, both in the United States and abroad, is profound. Society depends on massive, century-scale investments in infrastructure, including port facilities, commercial fishing, offshore energy, coastal development, municipal investments, national defense, and more. And, because the ocean is a major driver of climate, extreme events caused by the changing ocean threaten inland regions as well, whether it be via flooding, supply of food and energy, or other threads of the social tapestry. Likewise, when human communities depend on the ocean as a food source, for protection from storm damage, or for healthy shorelines, collapse of ecosystems due to extreme events can disrupt individual livelihoods and entire economies.

In light of humanity's dependence on the ocean, and the socioeconomic benefits of forecasting extreme events, improving the ability to forecast the occurrence of extreme events across spatial and temporal scales relevant to coastal communities is an urgent research priority.

Ocean-based extreme events fall into several broad categories (Table 2.2) and include phenomena acting over widely varying time and space. Specific examples of research pursuits to improve predictive capability include:

- Increasing the ability to model the future impact of global weather extremes, such as agricultural resilience and heat and flood preparedness.
- Forecasting the extent of deoxygenation and dead zones in the ocean, thus being able to anticipate their impacts on ecosystem health.
- Utilizing nature-based solutions that support societal adaptation and ecosystem restoration when forecasting responses to extreme events.
- Integrating existing and new data and knowledge from across disciplines and communities to improve the ability to respond to extreme events.
- Integrating existing and new data and knowledge to elucidate precursory signals that can improve forecasting of earthquakes, underwater volcanic eruptions, submarine landslides, and associated extreme events such as tsunamis.

TABLE 2.2 Categories of Ocean-Based Extreme Events

Geophysical/Geological Hazards	**Weather and Water-Related Hazards**	**Biological Hazards**
<ul><li>Earthquakes</li><li>Volcanism</li><li>Submarine landslides</li><li>Seabed instability</li><li>Tsunamis</li></ul>	<ul><li>Coastal storms and flooding, particularly in the context of sea level rise</li><li>Hurricanes</li><li>Heat waves</li><li>Atmospheric rivers</li></ul>	<ul><li>Eutrophication</li><li>Mass die-offs of fishes</li><li>Dead zones/deoxygenation</li><li>Harmful algal blooms or blooms of seaweed</li><li>Heat wave–induced coral bleaching</li><li>Urchin barrens and kelp forest depletion</li><li>Coastal pollution and maritime disasters</li><li>Overfishing</li></ul>

Societal Relevance and Broader Impacts of the Research

Earthquakes and Volcanism

The deep seafloor is generally out of sight and out of mind, but sudden movements associated with geological events can prove devastating to coastal communities, as the seafloor shifts abruptly. Earthquakes and submarine slides can generate tsunamis capable of killing tens of thousands or more people and animals, and inflicting billions of dollars of damage (Ranghieri and Ishiwatari, 2014). The largest earthquakes occur at subduction zones, which are frequently located close to land, and where 9+ magnitude events rip apart hundreds of miles of crust, generating shaking that can collapse buildings and tsunamis that can wipe out whole communities. The 2004 9.3-magnitude Sumatra earthquake generated a ~1,200-kilometer rupture with 30-meter-tall tsunami waves and a fatality estimate of over 200,000 human lives. Less than 7 years later, the Tohoku 9.1-magnitude earthquake generated 40-meter-tall tsunami waves, with an estimated 18,000 fatalities. The United States includes the Cascadia subduction zone, underlying Oregon and Washington, which is capable of generating magnitude 9+ earthquakes if it shifts, and the Alaska-Aleutian subduction zone, which generated one of the largest earthquakes ever recorded in 1964. It is certain that both of these subduction zones will produce earthquakes of devastating magnitudes and subsequent tsunamis again, but the timing is unknown. However, recent research has revealed precursory signals associated with some large earthquakes, leading to the goal of integrating data and models to improve real-time estimates of earthquake probabilities (e.g., Pritchard et al., 2020).

The 2022 Hunga Tonga–Hunga Ha'apai eruption in the south Pacific produced the largest underwater explosion ever recorded, spewing 146 metric megatons of water into the stratosphere (Jenkins et al., 2023) and about 900 kilotons of sulfur dioxide (Xia et al., 2024). This eruption was almost completely unexpected, illustrating how little is known about the processes occurring on the seafloor. The eruption and tsunami that followed resulted in six deaths; had it happened closer to a heavily populated area, it could have been much more devastating. Midocean ridge studies have revealed important information about seafloor eruptions and improved the ability to forecast them by revealing patterns of build-up seismicity and deformation thresholds, as well as noting other possible precursors (Nooner and Chadwick, 2016; Tan et al., 2016; Tolstoy et al., 2006; Wang et al., 2024; Wilcock et al., 2016). Nonetheless, the Hunga Tonga–Hunga Ha'apai eruption illustrates how much remains to learn.

Several major ongoing community efforts are working to design new observing systems and modeling frameworks to understand the limits and possibilities of geohazard forecasting at subduction zones, including those offshore Cascadia and Alaska. Examples of these efforts include the Cascadia Region Earthquake Science Center (CRESENT), the Subduction Zones in 4 Dimensions Initiative (SZ4D), and the Ocean Observatories Initiative.

Extreme Weather

Both inland and on the coast, extreme weather events—such as heat waves, hurricanes, atmospheric rivers, and flooding—increasingly threaten lives, property, and economies. Heat waves fuel urban heat islands, endangering public health and catalyzing wildfires. Extended periods of extreme warming in the ocean devastate marine life, bleaching coral reefs and causing toxic algal blooms. Hurricanes bring winds, rains, and storm surges that can transform bustling coastlines. In 2022, three hurricanes in the United States caused over $1 billion in damage each (NOAA, 2025b). Hurricanes have generally been strengthening in intensity since the 1980s and, increasingly, are rapidly intensifying before making landfall (Klotzbach et al., 2022) and traveling more slowly or even stalling over land (Hall and Kossin, 2019), exacerbating flooding. The occurrence of exceptionally strong atmospheric rivers[3] has been increasing. These are capable of transporting water vapor at a rate equal to 7–15 times the average daily discharge of the Mississippi River (Payne et al., 2020; Ralph and Dettinger, 2011). Flooding remains the deadliest type of natural disaster (Han and Sharif, 2021), and coastal flooding is worsened by rising sea levels (including "sunny-day" flooding; Piecuch and Hamlington, 2023) and intensified storms (Moftakhari et al., 2017), undermining infrastructure and displacing communities.

The cumulative impact of these multiple extreme weather events around the United States is a crisis of unprecedented proportions. When infrastructure, from roads to power grids, is heavily strained, agricultural lands are desiccated or inundated, property values plummet, insurance coverage becomes cost prohibitive, and economies that rely on fisheries and tourism are disrupted. Most importantly, human lives are lost, displaced, or irrevocably altered.

---

[3] *Atmospheric rivers* are formed when large-scale weather patterns create narrow channels of moisture-laden air.

Studying these extreme weather events is imperative for several reasons. First, improved understanding of patterns and trends allows for better predictions and early warning systems. More accurate predictions and early warning systems provide a scientific basis for informed decision-making, including disaster preparedness, infrastructure investments, and land-use planning. They can also be a powerful tool for advancing environmental justice. Public officials and administrators can assess the vulnerabilities of communities that are economically disadvantaged, identify the root causes of this vulnerability, and develop targeted adaptation strategies. Last but not the least, additional support for research and education on extreme events could expand and diversify the ocean science workforce, leading to improved understanding of local vulnerabilities and priorities and effective decision-making processes.

Biological and Ecological Extreme Events

Ocean organisms, from microbes to alga and animals, can also exacerbate extreme events and, in some cases, even create them. One example is ocean deoxygenation, which is a reduction in the amount of dissolved oxygen in the ocean. This is, in large part, due to increasing water temperatures (warmer surface water holds less dissolved oxygen and is less amenable to mixing with deeper water), as well as increased nutrient inputs from anthropogenic sources, such as agricultural runoff (Box 2.4). Increased nutrients in coastal waters often lead to phytoplankton blooms, which result in massive increases in biomass that cannot be sustained. Upon the death of the phytoplankton, the biomass is decomposed by microbes that consume more oxygen, leading to widespread hypoxia or even anoxia (i.e., waters effectively devoid of oxygen). Such massive, widespread deoxygenation events pose a severe threat to marine species living in the region, especially those that are highly sensitive to low-oxygen environments, including some commercially valuable fish and invertebrate species (e.g., shrimp).

Hypoxic zones (Box 2.4) are most common in coastal waters, where nutrient inputs from anthropogenic activities are most concentrated. However, even naturally occurring oxygen-deficient zones (or ODZs)—which are highly productive regions of the ocean that have lower oxygen concentrations at deeper depths (Margolskee et al., 2019)—may be expanding in size at higher rates. This too has a marked impact on neighboring ecosystems, including those of relevance to human health (e.g., sustenance fisheries) and economic well-being (e.g., commercial fisheries). For example, the eastern tropical Pacific and the Arabian Sea ODZs have expanded in size in recent years and impacted regional marine life and human populations that depend on these ecosystems for food and economic activities.

Increased temperatures, nutrient loading, and related factors have led to an increase in harmful algal blooms (HABs), also known as red tides. The impacts of HABs are not trivial, costing millions of dollars in lost revenue due to fishery and beach closures, and cleanup of dead biota (Sellner et al., 2003). HABs also contribute to deoxygenation zones previously noted and pose significant risks to marine life (e.g., domoic acid impacts on pinnipeds [McCabe et al., 2016]) and human health (CDC, 2024; U.S. EPA, 2024). Yet mechanistic understanding and synoptic coverage are insufficient for forecasting HAB events beyond few very isolated locales nationally (Davidson et al., 2016; Link et al., 2023). As global temperatures continue to rise and coastal pressures continue to increase, HABs are expected to become more frequent (Anderson et al., 2021), so the ability to forecast and mitigate them is important.

Forecasting capacity is not robust enough to estimate the magnitude and extent of ocean deoxygenation events. Likewise, other heat waves generate mass die-offs of coastal species such as corals, which can lead to algal dominance in nearshore reefs, reduced fish populations and reduction in fishing opportunities. As global temperatures continue to rise, the impacts of deoxygenation and heatwaves are expected to intensify, leading to further degradation of marine ecosystems (Oschlies, 2021). Improving the capacity to forecast and reduce impacts of ecological extreme events begins with integrating knowledge across communities, including the terrestrial and ocean science communities; coordinating international scientific efforts to determine how to mitigate greenhouse gas emissions; improving land-use practices; and providing scientific data that promote sustainable fisheries management.

BOX 2.4
The Gulf Coast Hypoxic Zone

Every summer, a low-oxygen (hypoxic) area develops off the Texas-Louisiana shelf when nutrient-laden fresh water from the Mississippi and Atchafalaya rivers flows from the interior United States into Gulf waters; this area is approximately 20,000 km^2, or roughly the size of Massachusetts (Figure 2.7). The Mississippi River and its tributaries meander more than 2,000 miles through parts of 32 U.S. states and Canada, collecting sediment, fresh water, and nutrients from the 1-million-square-mile drainage basin and bringing them southward to the coastline. This annual environmental event has been studied relentlessly since the mid-1980s, and numerous efforts have been made to mitigate and reduce the size of this zone, with little to no impact on its size but with increasing knowledge of how this and other hypoxic zones form and occur. Of note, the fishing community in the Gulf has long tracked the emergence and extent of the zone, and incorporated knowledge of its presence in fishing approaches.

The hypoxic zone off the U.S. Gulf Coast is a pertinent example of where a transdisciplinary research approach would be valuable, given the breadth of factors that influence development of this condition and the increasing occurrence of hypoxic zones in coastal areas globally. Solving this issue is complex because of the distance between the nutrient sources and the waters affected. Another challenge with this and other hypoxic zones is that synergistic effects of a changing climate converge with nutrient introductions, making scientific understanding and mitigation a moving target; this appears to the public as little progress with a long-standing problem. The enormity of the scale and scope of hypoxic zones transcends multiple jurisdictions, including international borders, and finding a solution transcends multiple disciplines. A transdisciplinary approach that melds ocean science with limnology, agricultural science, political science, education, industry knowledge and observations, and more is thus required.

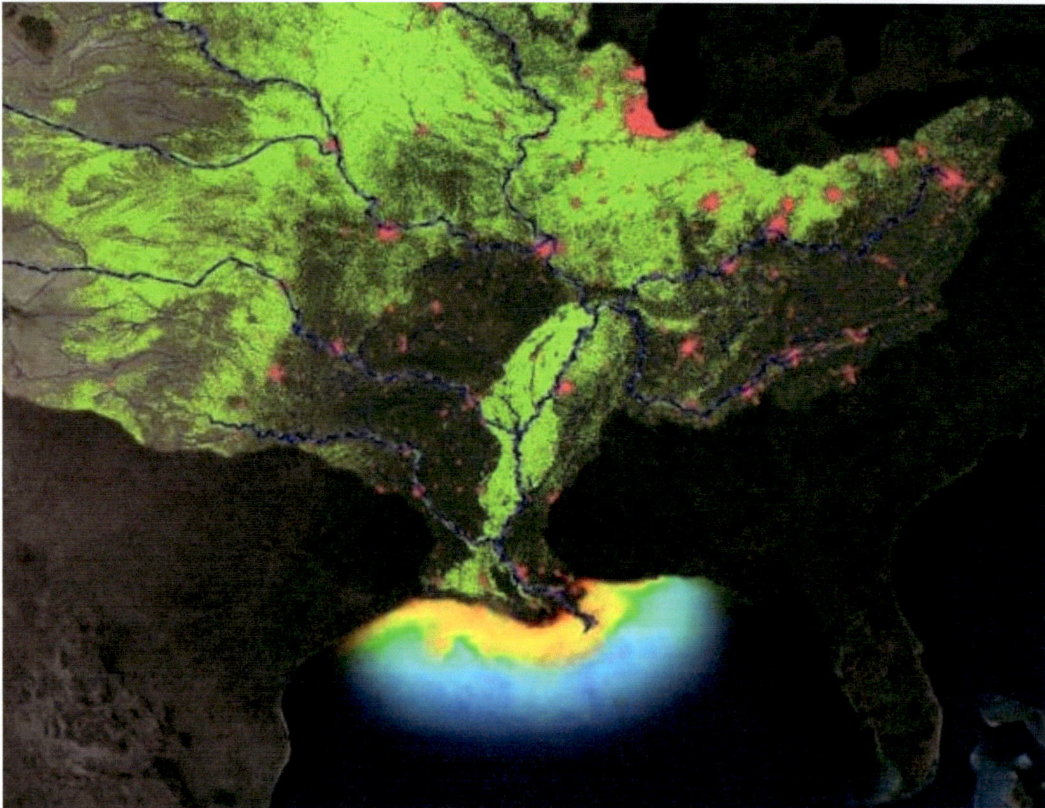

FIGURE 2.7 Hypoxic zone extending off shore from the Mississippi River basin.
NOTE: The map generally illustrates how runoff from farms (inland green areas) and cities (inland red areas) drain into the Mississippi River, bringing nutrient-rich water into the Mississippi basin, which causes an annual hypoxic event.
SOURCE: National Oceanic and Atmospheric Administration.

Many of the hazards discussed in this chapter are either caused or exacerbated by localized anthropogenic activities. Human activities, such as petroleum leaks and spills, can lead directly to extreme events that have a substantial impact on ecosystem and human health (NASEM, 2022c), coastal habitat destruction exacerbated or resulting from dredging or construction, and those discussed in the previous section on ecosystem resilience. The committee suggests that developing the means to predict the nature and extent of impacts on ecosystems and society will enable interest holders, including the scientific research community, to better mitigate any unintended events. Developing such a capacity requires better engagement among users (e.g., the scientific community, U.S. and Indigenous policy makers, and more), as well as continued integration of existing data and generation of additional data as needed.

Techniques, Tools, Technologies, and Human Resource Needs

Ocean-based extreme events described above include phenomena over widely varying time- and space scales. Researchers who study these events are united by shared reliance on basic ocean observations (data) that provides key scientific knowledge that will lead to enhanced forecasting capabilities, thus improving society's ability to prepare for and respond to extreme events. Notably, today such data are assimilated into next-generation quantitative and qualitative models with increasing sophistication enabled by advances in computer science and technology and, thus, computational oceanography and geodynamics. Enhancing the future predictive capacity of these models results from the dynamic feedback between model development, ongoing gathering of additional data to improve model performance, and model validation based on significant and intentional constraints on the "where" and "when" probability of extreme events. Better forecasting of ecological extremes like ocean deoxygenation and even extreme hydroclimate variability, such as droughts and extreme precipitation, will also benefit from "observations" provided by ocean archives of natural interannual and decadal climate oscillations. These archives integrate variability in large-scale ocean and atmospheric circulation as well as biogeochemistry, and thus provide the mechanistic insights required for predicting regional and global impacts. For biological extreme events, understanding the underlying physiological responses of species to extreme environments can be a major element of forecasting the impact of these events. In the case of biological responses, genetic adaptation to environmental variation can lead to shifts in species' responses (DeFilipplo et al., 2022; McManus et al., 2021). For these kinds of extreme events, an evolutionary adaptive element (ability to alter structure or habit for survival) is an essential element of forecasting.

Understanding the nuances of extreme events necessitates a wide array of perspectives. The convergence of traditional ecological knowledge, local expertise, and scientific acumen leads to a comprehensive picture of risk. This convergence also helps to understand local vulnerabilities, priorities, and capacities. A transdisciplinary approach to doing science fosters a more holistic ocean science community.

Developing more accurate and timely early warning systems in light of increasing storm frequency and intensity will rely both on increasing ocean observations of temperature and upper-ocean characteristics, as well as models that are able to nimbly use the data in near-real time in order to inform coastal municipalities about impending arrivals. While earthquake early warning systems, such as ShakeAlert, can provide seconds to minutes of notice before the arrival of damaging seismic waves after an earthquake has occurred, forecasts of when and where earthquakes will occur currently lack sufficient accuracy to be useful for short-term disaster preparations, such as evacuations, although they can inform building codes and other emergency preparedness. Sometimes, signals such as seismic foreshock sequences and slow slip transients precede dangerous earthquakes (e.g., Ito et al., 2013; Kato et al., 2012; Obara and Kato, 2016; Ruiz et al., 2014; Socquet et al., 2017), suggesting that once these phenomena are better understood, forecasts may be provided sufficiently far in advance to enable threatened population centers to better prepare. Furthermore, earthquakes are often associated with other damaging geohazards, such as tsunamis and landslides, and understanding the interrelations between these phenomena will help communities prepare and respond to future natural disasters. Early warning capability can also be enhanced through real-time offshore instrumentation that can improve the warning times by tens of seconds and the accuracy of expected shaking, as well as confirm the generation of tsunamis.

Scientific ocean drilling contributes significantly to better understanding extreme events on the seafloor such as earthquakes and submarine volcanism. Placing instruments, such as seismometers, in seafloor boreholes enables measurements of seismic events with much higher sensitivity than measurements from instruments placed on the seafloor surface (see Figure 3.14 in NASEM, 2024b). For example, borehole instruments placed in strategic locations near the Nankai trench off the coast of Japan led to key advances in understanding the fault systems of subduction zones. These observatories are an order of magnitude more sensitive to fault slip than other approaches (e.g., seabed observations), allowing smaller events to be detected and assimilated into numerical models and improving potential for future earthquake forecasting. The results of this research could have direct societal benefit, enabling better preparation and mitigation of future geohazard risks. Connecting borehole instruments to cabled seafloor observatories, as has been done along the coast of Japan, provides real-time assessments of fault slip transients, which precede earthquakes; this could be replicated in the United States.

3

Opportunities and Strategies for Accelerating Progress in Ocean Science

In addition to developing "a concise portfolio of compelling, high-priority, scientific questions," the statement of task asked the committee to "identify opportunities and strategies to promote innovative multidisciplinary and multi-sectoral approaches" to address these science challenges (see Box 1.2 in Chapter 1). In response to this task, the committee developed a framework for accelerating progress in the field of ocean science rooted in transdisciplinary[1] approaches to research, workforce development, and partnerships. The committee champions the continued need for basic research and advocates for a transdisciplinary lens to be utilized in planning and advancing basic research, and applying the research results.

A TRANSDISCIPLINARY COLLABORATIVE FRAMEWORK FOR OCEAN SCIENCES

Natural changes that are physical, biological, geological, or chemical and actively affect the ocean are generated by a wide range of mechanisms and processes known collectively as ocean processes. Ocean processes are integral to the planet's living and life support systems. Ecosystems at all scales depend on or are affected by all ocean processes. This is also true for humans, who have a deep history of interaction with the ocean (Erlandson, 2001). Much of conventional science research has historically been dualistic—treating humans as fundamentally different and separate from the rest of nature—which can unintentionally limit the researcher's ability to connect scientific insights on the natural world to human health, society, and culture (Berkes, 2003; Ivakhiv, 2002). While the specific details of societal needs and the role of the ocean in supporting those needs can vary based on social, cultural, political, and historical context, at its core, the overarching objective of public policy is to support and improve the long-term well-being of humans and the planet's living and life support systems (e.g., Breslow et al., 2016).

A focus on human well-being and its relationship to ocean processes can provide an enduring connection that places the ocean sciences in important multidisciplinary (natural, social, and health sciences) and multisectoral conversations. This approach is key to supporting human well-being by emphasizing the value of ocean science research and solving the global ocean challenges being faced. As such, in the next decade, research that connects ocean processes to human well-being through systems thinking[2] and transdisciplinary approaches will constitute an important step towards overcoming the historical human–nature dualist perspective and also engage the field of ocean science with issues of social and cultural importance (Box 3.1).

The path toward conducting successful transdisciplinary ocean science research is partially laid, but it is not easily put into practice. Research areas for which transdisciplinary approaches have already been identified and partially implemented include ocean acidification, marine heat waves, and hypoxia (all discussed in Chapter 2; Yates et al., 2015). For instance, Yates and colleagues (2015) provide a case study of the importance of transdisciplinary science for achieving societally relevant outcomes in ocean acidification research, a chemistry problem that is deeply intertwined with ecosystems and society:

> The global nature of ocean acidification (OA) transcends habitats, ecosystems, regions, and science disciplines. The scientific community recognizes that the biggest challenge in improving understanding of how changing OA conditions affect ecosystems, and associated consequences for human soci-

---

[1] See Box 1.3 in Chapter 1 for the committee's definition of *transdisciplinary research.*
[2] *Systems thinking* means considering a system holistically, in this case including humans as a component of the ocean system.

ety, requires integration of experimental, observational, and modeling approaches from many disciplines over a wide range of temporal and spatial scales. Such transdisciplinary science is the next step in providing relevant, meaningful results and optimal guidance to policymakers and coastal managers (p. 213).

BOX 3.1
Transdisciplinary Research for All Communities

Ocean science research and problem-solving in support of human well-being needs to implement scientific approaches that help mitigate the adverse effects of climate change on communities, including those that are most vulnerable. For example, many Pacific Island nations are facing adverse effects of sea level rise, lack of fresh water, extreme storm events, drought, ocean acidification, saltwater intrusion into groundwater, and biodiversity loss (Ober and Waters, 2023). As many communities across the globe are on the front line of climate change, it is now crucial to seek solutions together, with various types of knowledge informing research, forecasting, and ultimately a global change adaptation plan (Figure 3.1; Adger et al., 2006; Jodoin et al., 2021; UN, n.d.).

FIGURE 3.1 A man carries his belongings to safety after a flood in Mozambique.
SOURCE: Per-Anders Pettersson.

Collaborative development of transdisciplinary research, including basic research, is essential to finding innovative solutions to the problems affecting communities on the front lines of climate change. Often, engaging in collaborative production of knowledge and building transdisciplinary teams is challenging, producing unsuccessful results. However, taking the time and effort to conduct and resource successful transdisciplinary research in the next decade is essential for solving these problems.

These collaborative efforts, however, have not yet enabled the full understanding of the complex interactions of multiple stressors on human-caused changes in the ocean and the response to those changes by marine ecosystems. To succeed in understanding these interactions, a culture shift is needed, bringing together multiple knowledge holders (e.g., oceanographers, biotechnologists, the modeling community, business leaders, data scientists, engineers, Indigenous and local knowledge holders, local interest holders, policy makers, educators) to conduct ocean science research and generate predictive models for forecasting ocean and ecosystem change (e.g., Cavaleri Gerhardinger et al., 2024; Garibay-Toussaint et al., 2024). The process of achieving true transdisciplinary and multisector knowledge-building will necessitate that collaborators share responsibility and harmonize priorities throughout the various stages of the research process, from conceptualization of knowledge needs and framing of research questions to information-gathering, interpretation, writing, and dissemination (Christie, 2011; Webster et al., 2022).

Leveraging Multiple Knowledge Systems to Foster Transdisciplinary Collaboration

Traditionally, innovation in ocean sciences has centered on natural scientists, engineers, and industry professionals and has marginalized social scientists, historians, tribal and Indigenous leaders, and community members. However, a transdisciplinary approach requires that everyone with a deep understanding of the topic area be included in the early stages of the research and innovation process. It requires recognizing experience and cultural knowledge as true expertise, on equal terms with formal education in traditional research science. This approach may also expand the way parts of a research project are conducted, moving from precise agendas, timelines, and milestones toward a more interactive, exploratory, discovery-based methodology for combining talents and knowledge (e.g., two-eyed seeing;[3] Reid et al., 2021). Furthermore, research teams with strong intercultural competency can help translate critical findings across communities, cultures, and languages, yielding more actionable outcomes. Examples of scientific research collaborations that include transdisciplinary approaches are the National Science Foundation (NSF)-funded Cascadia Coastlines and Peoples Hazards Research Hub[4] (2025), a project that incorporates a range of knowledge systems, fostering transdisciplinary science, with the goal of helping "Pacific Northwest coastal communities prepare and adapt to coastal hazards through research and community engagement" (para. 1; see Box 3.2). These programs engage social scientists, modelers, regulatory agencies, educators, and cultural practitioners to explore solutions to pressing issues. As such, through transdisciplinary collaboration, advancements have been made to further an understanding of Earth systems.

As funding organizations seek to support collaborative science, with an emphasis often placed on broader impacts and meaningful community engagement, resources outlining best practices and strategies for achieving meaningful and successful engagement with nonacademic groups are needed. This requires developing collaborative processes that allow research focuses to expand to include community concerns such as well-being, cultural values, and livelihoods, as well as emergent impacts of research activities on local social dynamics. Some guidelines currently exist (Kūlana Noiʻi Working Group, 2021; NOAA, 2022; Protection of the Arctic Marine Environment [PAME], 2019), yet efforts to outline a set of guidelines would promote the success of future transdisciplinary projects.

Systemic Change in Valuation of Transdisciplinary Research

Conducting successful transdisciplinary ocean science research will take dedicated time, effort, and a more collaborative mindset during project inception, development, and output. Transdisciplinary science includes valuing and championing reciprocal relationship-building, implementing dedicated strategies for partnerships, and conceptualizing use-inspired research that includes multiple discipline and knowledge systems. True and effective knowledge co-production requires democratization and decolonization of knowledge-sharing structures. For example, people are more likely to share treasured, cultural knowledge and participate in its application in the research project if they have confidence that their contribution is valued (García-Quijano and Pizzini, 2015). This requires a commitment to cooperative formats and venues that empower oral history and knowledge-sharing rather than asymmetrical formats such as conference presentations or public hearings, which can favor those who command technological tools such as computers, statistical-graphing software, and presentation software.

One challenge to fostering collaboration across knowledge systems and disciplines is rooted in the way academic scientists, funded by NSF and other agencies, are assessed in decisions about attribution, hiring, tenure, and promotion. Many current assessment efforts consider metrics that can be quantified easily, such as number and amount of funded grants, and number and citations of publications; however,

---

[3] Viewing the world through different perspectives (often referring to combining Indigenous and Western views) to create a new way of seeing.

[4] See https://cascadiacopeshub.org.

BOX 3.2
Embracing the Spirit of Cascadia

The Cascadia Coastlines and Peoples (CoPes) Hazards Research Hub encompasses teams of researchers working to improve forecasting and coastal community resilience against natural hazards with a focus on the Pacific Northwest.[a] The Cascadia coastline is coincident with the Cascadia subduction zone, which extends from Cape Mendocino, California, to Vancouver Island, Canada. Coastal communities in this region face an array of risks, from decadal-scale rising sea levels and changes in weather, to abrupt hazards caused by ground-shaking, landslides, and tsunamis driven by subduction zone earthquakes. Changes in relative sea level along the coast are further exacerbated by tectonic motions associated with the subduction system.

The CoPes Hub brings together scientists and local communities to make transformative improvements to the understanding of coastal hazards (Figure 3.2). Coastal communities in Cascadia boast a wealth of cultural, social, and governance histories, along with traditional and local ecological knowledge. Their identities, values, and economies are deeply rooted in their coastal environments and ecosystems. Through community engagement and co-production, the CoPes Hub seeks to improve the ability of these communities to adapt to coastal geohazards.

FIGURE 3.2 Workshop on Navigating Coastal Hazards, March 2024.
SOURCE: Alessandra Burgos, Oregon State University.

The CoPes Hub is organized around five teams: (1) tectonic geohazards; (2) inundation and coastal change hazards; (3) community adaptive capacity; (4) science, technology, engineering, arts, and math (STEAM) education; and (5) community engagement and co-production of coastal hazard knowledge. Work is focused in "collaboratories," allowing researchers to partner with local communities with varied geologic and ecologic characteristics. Pilot projects from outside the project team are funded annually to support emergent research ideas and engage local communities. Examples of ongoing pilot projects include building multihazard evacuation map prototypes for coastal communities and establishing a community-driven earthquake monitoring program at the Quileute Tribal School. The CoPes Hub also facilitates opportunities for geoscientists to engage with coastal planners, community leaders, emergency managers, and state and federal agencies to discuss how geohazard research can be tailored to meet coastal community needs.

[a] The CoPes Hub is funded by the National Science Foundation's Geosciences Division of Research, Innovation, Synergies, and Education.

systemic racial disparities affect funding rates at NSF and likely other research funding bodies (Chen et al., 2022), promoting cascading impacts that perpetuate bias in the types of research conducted, decrease geographical distribution of funds, and negatively impact transdisciplinary workforce development. A related challenge is the lack of appreciation by academic administrators for the extra effort required to promote meaningful engagement of local communities and Indigenous groups, which can also affect publication output and related criteria for promotion. Even though many scientists conduct societally relevant research, few assessments exist that focus on quantifying the ways in which a research project contributes to the greater societal good. Systemic change in evaluation that encourages using all-embracing research practices, including initiatives that promote societal benefit, will enable fulfilment of the vital research goals outlined in this report. These revised evaluation criteria may include fair considerations of authorship and attribution; data management and access; data translation and dissemination; data sharing; use of collective benefit, authority to control, responsibility, and ethics (CARE) principles and findable, accessible, interoperable, and reusable (FAIR) principles; and application of metadata labels to protect Indigenous and local knowledge, such as traditional knowledge labels used by the Local Contexts Project.[5]

Practical Steps to Building Transdisciplinary Ocean Science Research Teams

Building transdisciplinary research teams is a path towards increasing broad participation in science, technology, engineering, and mathematics (STEM), promoting co-production with communities, and solving problems that include many interdependent factors and processes. Those conducting ocean science research must first recognize that a collaborative culture is a people-first structure within the overall research enterprise (Sahneh et al., 2021). Building this collaborative culture for successful transdisciplinary ocean science research teams will take time and dedication to (1) establish shared understanding of complex scientific problems and develop shared research goals, (2) co-develop conceptual frameworks and research designs that integrate approaches from multiple fields and perspectives, (3) execute the planned research, and (4) translate research findings to innovative solutions for marine ecosystems and human well-being. For example, the transdisciplinary teams needed to address pressing issues related to the three priority themes emphasized in this report—ocean and climate, ecosystem resilience, and extreme events—are all different and may leverage experts from the computing and technical modeling community, as well as place-based knowledge holders such as subsistence fishers to refine models and research designs.

Environmental and conservation research communities provide many recommendations for avoiding last-minute, shallow engagement with local and transdisciplinary research partners. These include enabling trust-building via low-budget, limited-scope initial projects; lowering barriers to including local partners in the grant application process (Rayadin and Burivalova, 2022); and applying multiple methods of co-producer engagement, such as workshops, community partner advisory groups, participatory mapping, and scenario development (Kliskey et al., 2023). An underutilized resource is university-based extension services that have a mission to translate research-based findings, good practices, and information to address interest-holder needs, including state, industry, and community needs. Extension professionals have long and trusted relationships with the communities that they serve. Their work, when adopted, contributes to the democratic process of knowledge generation as communities and businesses work hand in hand with students and researchers in academia to address local issues. The National Sea Grant College Program, created in 1966 and originally housed in NSF, combines research, education, and outreach to address coastal issues of resource management, academia–industry interactions, environmental quality, and economic competitiveness. Programs such as Sea Grant, as well as community-based private organizations such as The Nature Conservancy or Conservation International, can be leveraged to build trust and engage local communities in transdisciplinary research.

Table 3.1 outlines a framework for building transdisciplinary teams that increases collaboration and aims to address complex scientific challenges and improve human well-being (Hall et al., 2012). NSF's Division of Ocean Sciences (OCE) could adapt the four-phase model approach, originally applied to the

---

[5] See https://localcontexts.org/.

health sciences, to offer as guidance to proposers on how to structure transdisciplinary ocean science research teams. Membership of such teams would necessarily include scientists from multiple disciplines and sectors, as well as partners from community groups, in various phases of the research process. Additional research on this topic emphasizes the importance of managing communication among and ensuring accountability of all team members (Waite et al., 2023). Other resources include models of different leadership approaches for small and co-located transdisciplinary teams versus large and disperse teams (Gray, 2008), and insights on how to facilitate and distribute leadership within a large transdisciplinary team to improve collaboration and support transdisciplinary research objectives (Archibald et al., 2023). Additionally, social network analysis offers quantitative and qualitative indicators important in the assessment and evaluation of transdisciplinary research team effectiveness (e.g., Steelman et al., 2021) and for identifying influential actors and information brokers, as well as barriers and opportunities in collaborative networks (Hoelting et al., 2014; Smythe et al., 2014). Lastly, transdisciplinary design features specifically applied to ocean development–financed projects in underresourced settings found that community participation and multiknowledge systems were key to progress (Hills and Maharaj, 2023).

TABLE 3.1 Four Phases of a Transdisciplinary (TD) Team-Based Research Model

	Developmental	**Conceptual**	**Implementation**	**Translational**
Primary Goal	Establish a shared understanding of the scientific or societal problem space of interest—including what concepts fall inside and outside its boundaries—and mission of the group	Develop novel research questions or hypotheses, a conceptual framework, and a research design that integrate and extend approaches from multiple disciplines and fields	Launch, conduct, and refine the planned TD research	Apply research findings to advance progress toward developing innovative solutions to real-world problems, as appropriate and to the level of science at which the research is conducted
Team Type(s)	• Network • Working group • Advisory group • Emerging team	• Emerging team • Evolving team	• Real team	• Adapted team • New team
Key Team Processes	• Generate a shared mission and goals • Develop critical awareness • Externalize group cognition • Develop a group environment of psychological safety	• Create a shared mental model • Generate shared language • Develop compilational transactive memory • Develop a team TD orientation	• Develop compositional, taskwork, and teamwork transactive memory • Conflict management • Team learning	• Adapt the team, as needed, to address translational opportunities • Generate shared goals for the translational endeavor • Develop shared understandings of how these goals will be pursued

SOURCE: Hall et al., 2012.

CONCLUSION 3.1: Transdisciplinary mindsets, skillsets, and practices—including collaborations with other disciplines such as humanities and social scientists, as well as stewards of Tribal, Indigenous, and other knowledge holders—will be the key to actualizing a new paradigm of ocean science. Multiple perspectives, not only for ocean science researchers but also among ocean science funding agencies and directorates, will advance innovation in applying basic research. Resource investment is needed, both fiscally and programmatically (including in leadership development, management, and outcome assessments), to support and incentivize transdisciplinary research teams to be created and sustained in the field of ocean science.

RECOMMENDATION 3.1: To foster transdisciplinary research that promotes emerging solutions to challenges related to changes in ocean systems and processes, the National Science Foundation's (NSF) Division of Ocean Sciences should invest in projects that utilize a participatory process toward relationship-building and collaborative efforts, establishing long-term trust and knowledge-sharing and impacting broad interests. NSF should explicitly enable research that crosses directorates and programs and intentionally facilitate efforts to dismantle barriers to transdisciplinary approaches and implementation. Potential strategies include:

- **supporting projects with measurable social and environmental returns on investments, especially use-inspired and solutions-oriented research on the changing ocean;**
- **investing in training an expanded ocean science workforce that includes developing essential science, technology, engineering, and mathematics skills as well as transdisciplinary skills that enable meaningful connections to the humanities, social sciences, and economics;**
- **implementing requirements for proposals that encourage partnerships with interest holders including local communities, regional organizations, Indigenous and Tribal groups, and other users, to participate in ocean research; and**
- **financially supporting platforms and networks that facilitate knowledge exchange between interest holders, sectors, and disciplines.**

WORKFORCE DEVELOPMENT

Coastal interest holders, particularly historically marginalized communities and underrepresented groups, face daily challenges in comprehending and navigating the intricacies of ocean and coastal governance. Gaining competencies to influence decision-making related to the ocean and coasts necessitates knowledge of historical social and ecological baselines, which are often obscured by limited awareness of marine ecosystems and gaps in intergenerational knowledge transmission. To effectively bridge the imbalances in ocean science capacity and promote public engagement in research, the ocean science enterprise needs to foster strategies and develop competencies that empower ocean stewards in decision-making. It requires including both scientists and nonscientists, especially from historically marginalized communities and underrepresented groups in ocean science research, education, and governance. Gerhardinger et al. (2024) suggests using a combination of marine learning networks (Bayliss-Brown et al., 2020; Dalton et al., 2020) with media and information ocean literacy strategies (Singh et al., 2016) to develop transdisciplinary capacities and engage nonacademic interest holders in co-production of knowledge, thus democratizing the venues and mechanisms for knowledge creation, access, and sharing, with the reconsidered experts as described above, with fair distribution across the globe. Building transdisciplinary capacities also requires sustained commitment and collaboration across multiple disciplines, sectors (public, private, academic) and scales (local, regional, national, international). Incorporating such an approach to ocean science research is important for training the next generation of the ocean science workforce. Box 3.3 describes one program aimed at increasing traditional, local knowledge in ocean sciences and the value this program has added to science, specifically to fisheries and marine science management in Alaska.

A wide array of perspectives and experiences in the ocean sciences workforce will contribute to innovation and is critical to meeting society's needs. People with different backgrounds, perspectives, and experiences working together leads to more innovative and creative problem-solving (Aminpour et al., 2021). Heterogenous teams can identify a wider range of potential issues and solutions than homogenous teams (Bendor and Page, 2018), and manuscripts from ethnically variant publishing groups have been found to be more impactful than those from homogenous groups (Freeman et al., 2014).

Furthermore, as discussed in the section on transdisciplinary research, broad representation in research improves the relevance of ocean science outcomes. Many coastal and island communities, with deep-rooted, historical–ecological knowledge about their marine environment, are on the front line of experiencing ocean

---

BOX 3.3
"All of Us": The Tamamta Program

FIGURE 3.3 Tamamta Fellow Elizabeth Mik'aq Lindley (Yup'ik) holding a sayak (or sockeye) on the Kuskokwim River, 2017.
SOURCE: Orutsararmiut Traditional Native Council, photo courtesy Elizabeth Lindley.

The National Science Foundation Research Traineeship Tamamta program at the University of Alaska Fairbanks uses Indigenous approaches to transform graduate education programs in fisheries and marine science fields, which tend to lack cultural representation in academia and the workforce (Figure 3.3). Named for a Sugpiaq and Yup'ik word meaning "all of us," the Tamamta program broadens graduate training to engage Indigenous students, to center Indigenous knowledge systems in current and new curricula, and to reach widely across the university and partner organizations toward larger system change.

As environmental and social challenges intensify in the Arctic, the need for bold and transformative action that crosses cultural, racial, and disciplinary boundaries grows. To respond to this need, Tamamta supports comprehensive graduate training that advances science, bridging Western and Indigenous knowledge systems for M.S. and Ph.D. trainees. Program activities include new team-taught interdisciplinary courses, an elder-in-residence program, a visiting Indigenous scholars' program, a fish camp and other cultural immersion experiences, professional development and cultural competency skill-building with faculty and agency partners, retreats, internships, coastal research experiences, hosted dialogues, and art installations.

Tamamta is dedicated to effective training of science, technology, engineering, and mathematics graduate students in high-priority interdisciplinary and convergent research areas through comprehensive traineeship models that are innovative, evidence-based, and aligned with changing workforce and research needs. The program has reached more than 100 students over the past 5 years and engaged federal, state, and tribal partner organizations toward larger system change within fisheries and marine science and management in Alaska.

---

changes and yet are underrepresented in ocean science research careers (Kitolelei et al., 2022; Thornton and Scheer, 2012). Increasing collaboration with these communities in ocean research sectors can ensure their needs and knowledge are incorporated, enhancing understanding and making the research more relevant and impactful for all (Fischer et al., 2015; Thornton and Scheer, 2012), while supporting transformative, transdisciplinary and multi-sectoral research that addresses global societal needs and human well-being (AlShebli et al., 2018).

There is a crucial need for education and training to develop transdisciplinary mindsets, skills, and leadership abilities among those conducting research. For transdisciplinary ocean science research to truly advance, current and future ocean science researchers need not only foundational disciplinary knowledge, but also additional knowledge and skillsets such as cross-cultural communication, emotional intelligence, and active listening skills, among others. For example, integration of knowledge and expertise across disciplines and communities is identified as a core challenge and the defining characteristic of transdisciplinary research (Hoffmann et al., 2022), but the traditional, siloed culture in higher education has been slow to transition to more integrated research approaches. There is some evidence of growing momentum toward facilitating greater transdisciplinary habits of mind and skills in undergraduate (Bammer et al., 2023) and graduate programs (Box 3.4; Horn et al., 2024; Kluger and Bartzke, 2020), including some work in marine research management (Ciannelli et al., 2014) and marine science (Wilson et al., 2021). Funding agencies can further catalyze change in ocean science education and research by investing in projects that support and assess the success of transdisciplinary approaches in undergraduate and graduate education and the transfer of knowledge from place-based knowledge holders.

The Current State of the Ocean Science Workforce

Current demographic trends in the ocean science workforce reflect a mixture of opportunities and challenges for increasing representation. For example, the percentage of non-White individuals in ocean sciences is low compared with the U.S. population. While new enrollment of non-White individuals in ocean science programs is increasing, that same trend is not seen in graduating students, indicating work is needed to retain these students (Figure 3.5a; Lewis et al., 2023; concerning contributing factors, see Graham et al., 2023). In addition, data suggest an overall decline in the number of U.S. citizens receiving degrees in ocean science disciplines (Lewis et al., 2023). As for trends in gender in the academic workforce (Figure 3.5b), data from 2007–2021 show ratios of women to men that are more or less equal in recruitment and retention of students; however, academic positions, particularly tenure-track and tenured, remain skewed toward employment of men.

Historically, among the Earth science disciplines, the ocean (and atmospheric) science workforce has notably low representation. The current lack of historically excluded representation may limit the breadth of perspectives available to tackle multifaceted ocean issues (Kappel et al., 2023). Previous National Academies reports (e.g., NASEM, 2023) have described the need to re-tool STEM education and build up a robust STEM workforce. Workers, including researchers, need both disciplinary and interdisciplinary knowledge and skills, and the ability and capacity to collaborate within transdisciplinary projects. This need increases the competition for obtaining skilled workers, significantly impacting hiring and retention issues. To ensure success in conducting basic ocean science research with transdisciplinary teams, the preparation and valuation of a skilled and innovative ocean science workforce is imperative.

Globally, most coastlines border low- and middle-income countries with less research and infrastructure capacity to address their nations' ocean resource and marine hazard challenges. This may also be the case for some parts of the United States (i.e., the Gulf Coast), a topic examined in Chapter 4. Additionally, many biodiversity hotspots exist in relatively understudied geographies nearby underresourced populations (Stefanoudis et al., 2021). Such regions may fall victim to *parachute science*, wherein scientists from wealthier countries or regions conduct field research without meaningful engagement of and benefit to the local community or region (de Vos, 2020; de Vos and Schwartz, 2022). The United Nations (UN) has declared 2021–2030 the Decade of Ocean Sciences for Sustainable Development,[6] recognizing the need for broader scientific engagement to create transformative ocean science solutions for sustainable development and societal needs. A shared framework called Leave No One Behind[7] is one result of the UN decade provided to address barriers and biases in ocean science research.

The majority of coastal communities worldwide currently lack the resources and capacity to monitor and understand their own waters. In places where Indigenous knowledge exists, it is often devalued and disregarded, such that there are unequal and unbalanced collaborations. Since the global pandemic in 2020, a plethora of publications has suggested ways to broaden participation, promote welcoming spaces, and advance social impartiality in ocean sciences (e.g., Ali et al., 2021; Crosman et al., 2022; de Vos et al., 2023; Kaikkonen et al., 2024; Kappel et al., 2023).

---

[6] See https://oceandecade.org.
[7] See https://unsdg.un.org/2030-agenda/universal-values/leave-no-one-behind.

BOX 3.4
Workforce Development and the Blue Economy

The impact of climate change on coastal communities—including sea level rise, ocean acidification, food security, and human health—represents a major challenge to local, regional, and global economies and industries. With greater societal awareness of environmental issues, demand is growing for sustainable economic and environmental programs, plans, and budgets at the local, regional, national, and global levels. In turn, demand is also increasing for ocean engineering, cyberinfrastructure, data science, technology, and management programs that specifically address workforce development and leadership training needs to support this blue economy.

Leadership programs are offered at the graduate level, such as the Blue MBA at the University of Rhode Island (Moran, 2021; Moran et al., 2009) and the Ocean Enterprise Entrepreneurship Program at the University of Southern Mississippi, which combine business management with ocean science training for students to develop skills applicable to the new blue economy. Some undergraduate programs (e.g., University of New England's B.S. in marine entrepreneurship) are also emerging. These programs align with recent programs and initiatives of the National Science Foundation's Directorate for Technology, Innovation and Partnerships, including the Networked Blue Economy and Convergence Accelerator programs, which are designed to tackle challenges related to climate, sustainability, food, energy, pollution, and the economy.

Considering the broad spectrum of challenges of, and opportunities for, the growing blue economy, there will be an increased demand for such transdisciplinary technical skills and advanced academic programs in fields such as fisheries and aquaculture, ocean technology, engineering, ocean observing, renewable energy (Figure 3.4), marine biotechnology, finance, and risk management.

Transdisciplinary research and socioeconomic monitoring of blue economy initiatives will play an important role in developing policies for an ecologically and socially sustainable blue economy that minimizes common pitfalls of economic development. These pitfalls can include increased social disparity, "ocean grabs" with displacement of local residents, and degradation of local marine resources (see Germond Duret et al., 2023; Narchi et al., 2024). The development of new and innovative instructional programs provides an opportunity for greater collaboration between industry, government agencies, and academia in building a workforce pipeline that meets the needs of the blue economy.

FIGURE 3.4 Offshore wind project off Virginia Beach.
SOURCE: Stephen Boutwell; Bureau of Ocean Energy Management.

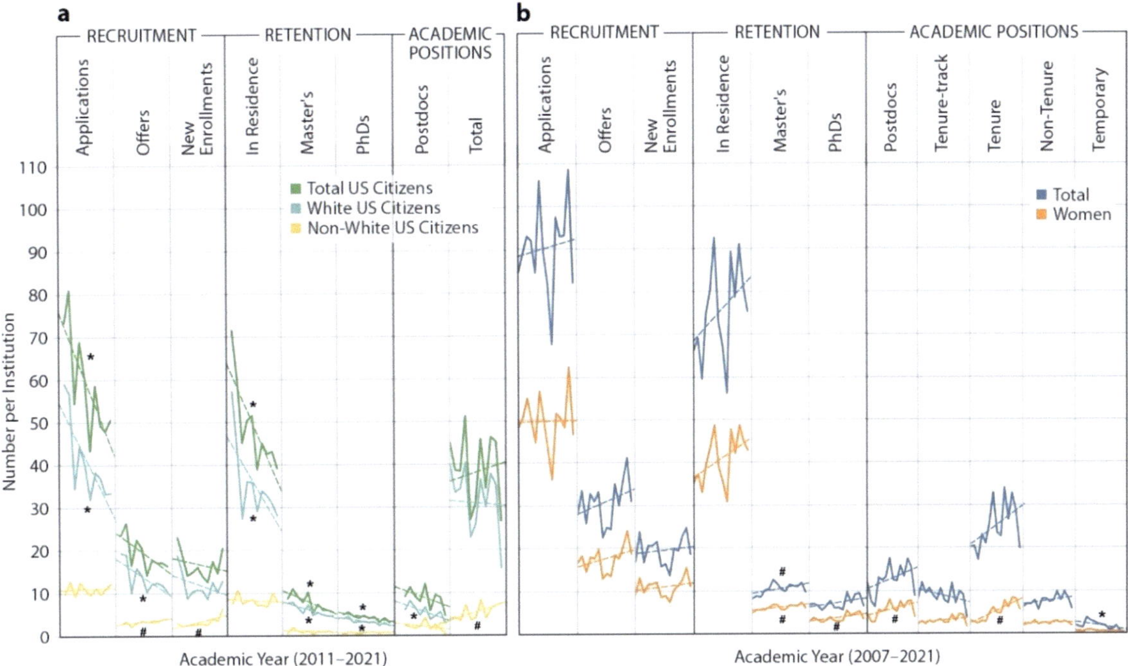

FIGURE 3.5 Trends in graduate students and academic positions in ocean science.
SOURCE: Lewis et al., 2023.

Another important issue that the ocean science enterprise needs to tackle is improving work environments. Geoscience disciplines struggle with bullying, hostile work environments, discrimination, and sexual harassment aboard research vessels and in the field (Harris et al., 2022; Marin-Spiotta et al., 2020; NASEM, 2018; Women in Ocean Sciences Report, 2021). Because of lacking codes of conduct, reporting mechanisms, and fear of retribution, inappropriate behaviors often remain unchallenged (Seafarers International Research Centre, 2003). Environments that normalize, deal ineffectively with, or promote sexual or other forms of harassment negatively impact the physical and mental health of people who are experiencing it as well as those of bystanders. Toxic or unsafe working environments also deter people from entering or remaining in the field of ocean science.

CONCLUSION 3.2: Within the field of ocean science, it is imperative to ensure that people from all career stages, backgrounds, and ethnicities are protected from sexual harassment and bullying.

Cultivating the Future Ocean Science Workforce

Developing the next generation of ocean science experts, engineers, practitioners, managers, and decision-makers requires a multipronged approach to education, research, mentorship, advocacy, and engagement.

Building Basic Skills for Emerging Technologies

Building up basic skills within the future ocean science workforce is imperative to success. The field of ocean science generates vast amounts of data from satellites, underwater sensors, and at-sea research expeditions. Thus, supporting the advancement of expertise in integrating emerging technologies such as artificial intelligence (AI) and machine learning into ocean science research not only helps analyze and interpret the data but also helps to improve the accuracy of oceanographic models and predictions. The skilled use of AI-powered systems could enhance ocean monitoring and surveillance efforts, for example, by detecting illegal fishing activities, monitoring marine habitats, or identifying environmental threats such as pollution or coral bleaching, with a low amount of manual labor. This type of expertise can also be used in sophisticated analyses of large datasets, such as genomic data and bioinformatics, which can help scientists understand and manage the impacts of extreme weather and climate change on ecosystem processes and biodiversity. Training the ocean science workforce in using and applying new and emerging technologies is imperative to advance understanding of the ocean, address environmental issues quickly and efficiently, and promote sustainable management of marine resources. The global and national networks of coastal marine laboratories play an important role in training the next generation of undergraduate and graduate students, as further described in Box 3.5. Among the 99 labs that are members of the National Association of Marine Labs (NAML),[8] 80 percent are associated with one or more universities and most offer an array of courses and community programs at all ages and levels of education.

BOX 3.5
Marine Laboratories as Training Ground for Emerging Ocean Scientists

Marine laboratories are centers for training ocean science undergraduate and graduate students; they often have coastal access and/or fleets of small boats that can sample nearshore marine environments. Through initiatives and opportunities to enhance transdisciplinary research, marine labs need to co-develop questions to help address community-driven research needs. Parachute science has been common, especially in areas near marine labs, such as in the tropics, where there is a large discrepancy between the academic scientists and the local communities (e.g., Stefanoudis et al., 2021). Thus, as marine labs are typically in remote locations, where Indigenous and local knowledge of native species, ecosystems, atmospheric sciences, and oceanography may still exist, collaborations should be encouraged in engaging with local communities and training the next generation of ocean scientists.

Many labs engage in foundational fellowships and programs to enhance student populations, broadening perspectives in the marine sciences fields (e.g., the National Science Foundation [NSF] Advanced Technological Education program targeting Pacific Islanders at the University of Hawai'i at Mānoa; Figure 3.6) and broadening geographic distribution, creating more balanced access points for entry into the field of ocean science than is currently offered by NSF-funded major ocean facilities (see Chapter 4). In 2023, the National Association of Marine Laboratories (NAML) reported that 435 internships were filled by underrepresented students across the member labs. NAML labs reported serving 199,592 K–12 students per year, as well as 9,572 undergraduates, 2,164 graduate students, and 2,707 teachers (NAML, 2023). They have also hosted summer camps and early childhood outreach and field trip programs for community engagement. U.S. marine labs are a significant influencer in educational experiences among the ocean science research infrastructure.

FIGURE 3.6 Participants in the Partnership for Advanced Marine and Environmental Science Training for Pacific Islanders.
SOURCE: National Science Foundation (see http://nsf-ate.pbrc.hawaii.edu).

---

[8] See https://naml.org.

CONCLUSION 3.3: Successfully addressing ocean science research challenges in the coming decade will depend on the cultivation and development of an expanded and transdisciplinary ocean science workforce. The workforce will benefit from receiving training in collaborative science methodologies, best practices for engaging in use-inspired research, and conducting research projects that meet societal needs.

Broadening Perspectives in the Ocean Science Workforce

Targeting recruitment for programs and fellowships from underserved communities, including Indigenous and local communities in the place of research (e.g., tribal colleges, organizations), redistributes investment within marginalized populations. Thoughtful and selective processes for the evaluation of qualified candidates for these opportunities will promote widening of perspectives and experiences within the field. Other strategies for broadening perspectives and place-based knowledge include providing training for teachers and principal investigators on collaborative science methods and multiple knowledge systems integration. Most important, a transdisciplinary lens requires working intentionally with the people of the place where research is being conducted, who have place-based knowledge of the ocean, coasts, and marine inhabitants, to seek unique solutions to pressing issues. It is also important to establish processes and guidelines to ensure a safe and welcoming culture in order to avoid tokenism and systemic bias. In addition, recruiting, training, and retaining international talent is crucial for the ocean science workforce and essential for maintaining U.S. research leadership. Incorporating international talent brings a multitude of perspectives, expertise, and innovation vital for addressing global marine challenges. Furthermore, researchers from less-resourced nations that are based at U.S. institutions are well-positioned to foster novel international collaborations that benefit both the United States and their home nations (when burdensome pitfalls are avoided; see Mwampamba et al., 2022).

Remote and autonomous sensors can provide global data coverage of the ocean, but rigorous, context-informed, and solutions-oriented science requires a global cadre of skilled researchers and collaborations. To ensure sustainable international collaborations, it is important to provide training for the U.S. workforce that fosters cultural competence, collaborative skills, ethical practices, and an understanding of differences in scientific approaches. Travel opportunities may also allow U.S.-based students and researchers to learn from established transdisciplinary ocean research programs in other countries. Partnerships for Enhanced Engagement in Research (PEER) was an international grants program established to support scientists in low- and middle-income countries that were partnered with U.S.-based collaborators on capacity-strengthening activities along the whole discovery-to-applied research spectrum. PEER was administered by the National Academies from 2011 to 2024 and managed by the U.S. Agency for International Development, with support from NSF and other federal agencies. Following the conclusion of funding for this program, NSF has the opportunity to support increased and more intentional collaborations between U.S.-based researchers and those in low- and middle-income global communities, enabling a fuller understanding of ocean sciences and supporting more international partnerships.

CONCLUSION 3.4: The field of ocean sciences represents an opportunity to engage the future and current workforce in an exciting area of expertise that will help develop STEM skills essential to a strong national workforce, regardless of whether the individual stays with the academic ocean sciences or ocean sciences more broadly. Every opportunity to go sea, work remotely with robotic systems, or work on samples taken from the water column or from seafloor cores can inspire and change the course of a career.

RECOMMENDATION 3.2: The National Science Foundation's Division of Ocean Sciences (OCE) is uniquely positioned to shape the future of ocean sciences research and policy by cultivating a workforce that includes multiple skillsets and knowledge systems. OCE should:

- **Explicitly support reskilling and upskilling, as well as mentorship, of the academic ocean science workforce to better engage with industry, entrepreneurs, interest holders, and other**

partners, to promote leadership development in seagoing technologies, data management, cultural competencies, and educational and mentoring practices.

- **Support workforce development by investing in vocational and academic pathways, such as offering scholarships and apprenticeships; establishing fellowships with thoughtful considerations of metrics that support underserved student and professional populations, including students, researchers, or interest holders in the locations where the research is being conducted; and funding and incentivizing collaborative projects that bring together researchers from varied geographic and disciplinary backgrounds.**

- **Promote safe working environments by enforcing and incentivizing policies that protect people from discrimination, harassment, and bullying.**

STRATEGIC PARTNERSHIPS

The committee acknowledges NSF's efforts to cultivate partnerships between academia, industry, nonprofits, government, civil society, and other sectors to pursue transformative research, solve societal problems, drive economic progress, and build a future-ready workforce. NSF currently engages with other agencies and industry to achieve common goals and missions. It engages with industry to provide funding for companies to conduct scientific research and further develop the workforce they need. NSF also enters into cross-sector partnerships that help inform research and workforce directions, including with industry and other interagency partners. But NSF is not a mission agency, and its primary role is in funding basic research needed to develop tools that are then transitioned to mission agencies. For example, NSF funded the development of biogeochemical (BGC)-Argo floats through the National Ocean Partnership Program (NOPP) and subsequently has funded the deployment of 500 BGC-Argo floats in the global ocean through the Mid-Scale Infrastructure program.[9] Recently, NOPP awarded $24.3M to advance research in marine carbon dioxide removal, with funding provided by the National Oceanic and Atmospheric Administration (NOAA), the Department of Energy (DOE), the Office of Naval Research, NSF, and the ClimateWorks Foundation. This funding opportunity is the "first large-scale public and multi-partner investment of research specifically focused on a suite of marine carbon dioxide removal approaches" (Ocean Acidification Program [OAP], 2023, para. 2).

In this regard, the committee commends OCE for its efforts to partner with agencies that bolster and leverage ocean science research more broadly, namely NOAA (ships), the National Aeronautics and Space Administration (satellites), and related mission agencies (U.S. Geological Survey, Bureau of Ocean Energy Management [BOEM], DOE, Office of Naval Research). This has been accomplished through interagency cooperative agreements and working groups. The committee encourages OCE to strengthen such interagency cooperation through existing federal policies and mechanisms to leverage NSF-funded basic ocean science research with those mission agencies. Additionally, the National Ocean Policy exists and could be modified as needed through additional executive order(s) to enhance the broader federal ocean research enterprise to meet national needs.

OCE also has a history of working with other NSF directorates as well as the Office of Integrative Activities to broaden opportunities for ocean researchers. For example, OCE has collaborated with the Directorate for Biological Sciences since the 2000s in funding several coastal Long-Term Ecological Research sites, such as the Santa Barbara Coastal, California Current Ecosystem, and Moorea Coral Reef, and more recently, oceanic sites such as the Northern Gulf of Alaska and North East U.S. Shelf.[10] These sites will be critical for documenting physical, chemical, and biological change in the ocean over timescales necessary to understand ecosystem resilience and the impact of climate change on heat and carbon budgets in the ocean. The ocean research community has also been successful in working with the Office of Integrative Activities to obtain substantial new funding for establishing science and technology centers such as the Center for Coastal Margin and Prediction, Center for Microbial Oceanography, and Center for Dark Energy

---

[9] See https://www.nsf.gov/awardsearch/showAward?AWD_ID=1946578 (accessed 12/23/2024).

[10] See https://lternet.edu/network-organization/lter-a-history/ (accessed 12/23/2024).

Biosphere Investigations, in the past, and recently the Center for Chemical Currencies of a Microbial Planet.[11] These centers have been successful at broadening participation and expanding available resources through partnerships with national labs, industries, and other entities. The Directorate for Technology, Innovation and Partnerships (TIP) program on Accelerating Research Translation also has potential for impactful partnership with OCE as it is focused on supporting higher education institutions that seek to build capacity and infrastructure for translation of fundamental academic research into tangible solutions that benefit the public.

These, and other strategic partnerships will no doubt be both necessary and critical to accomplishing the entire research portfolio laid out in Chapter 2. The committee acknowledges that, while the path to successful partnerships is the more difficult, the "path less traveled" (Figure 3.7) leads to several opportunities to work with entities outside of OCE to advance the ocean science of the next decade.

FIGURE 3.7 The road to partnership is the road less traveled.
NOTES: Scientist and oceanographer Michael Freilich was director of the Earth Science Division in the Science Mission Directorate at the National Aeronautics and Space Administration (NASA) headquarters from 2006 until 2019. One of Dr. Freilich's greatest accomplishments at NASA was to forge long-term and enduring partnerships between NASA and multiple international space agencies. The road sign featured in this photo had very special meaning to him. He was a master at building strong partnerships, and the image encapsulates his deep understanding of the nature of the work required to take the turnoff to Partnership Road—the less-traveled road.
SOURCE: Michael Freilich.

---

[11] See https://new.nsf.gov/od/oia/ia/stc#active-centers-c98 (accessed 12/23/2024).

Subject Areas Ripe for New Partnerships

Several opportunities for partnerships between OCE (i.e., chemical oceanography, biological oceanography, physical oceanography, and marine geology and geophysics programs, as well as major facilities, ships, and the Ocean Observatories Initiative) and other NSF directorates, as well as opportunities for public–private partnerships exist that would help progress the urgent ocean science research portfolio put forward in this report. Opportunities include addressing infrastructure needs such as the topics of big data, AI and machine learning, ocean model development and prediction tools, ocean observing platforms, and sensor development (and related technology/engineering), as well as emerging use-inspired research directions in ocean energy and aquaculture research and development (e.g., Advanced Research Projects Agency-Energy Macroalgae Research Inspiring Novel Energy Resources), seabed mining for critical minerals, marine carbon dioxide removal, and subseafloor carbon capture. These inter- and transdisciplinary areas overlap with other federal agencies and organizations and represent an opportunity to leverage federal investment with the private sector. A few of these examples are highlighted in further detail below.

Ocean Observing

Several autonomous surface vessel companies have greatly expanded the capacity and capability to observe the ocean, with a suite of uncrewed platforms developed to collect ocean (e.g., temperature salinity, dissolved oxygen) and seafloor bathymetry data, among other ocean observing capabilities. These companies (e.g., Saildrone, SeaTrac, Liquid Robotics, OceanAero, Subsea Sail) may represent an opportunity for OCE-funded scientists and engineers to work with industry to utilize the technology or the data being collected. Additionally, the Naval Oceanographic Office maintains a fleet of gliders and other remote vehicles that could be made into a shared resource through a partnership with the Office of Naval Research (ONR). Traditionally, NSF and ONR have worked together to provide access to research vessels (see Chapter 4), but this partnership could be strengthened through joint projects, technical exchanges, and appropriate data sharing.

Fiber optic cables also present the opportunity to make high-precision measurements that can be used for seismic sensing. Specifically, distributed acoustic sensing is a relatively new technology that allows real-time measurements along the full length of a fiber optic cable. Additionally, telecommunication cables can be instrumented through the SMART (Science Monitoring and Reliable Telecommunications) cables initiative. Partnerships with telecommunication companies have the potential to vastly increase the area of the seafloor that can be monitored seismically, improving earthquake warning systems and seismic imaging of Earth's interior. For a relatively low cost, existing fiber optic cables can be instrumented for distributed acoustic sensing. Moreover, while traditional approaches previously necessitated using "dark" (or unused) fibers, new technologies are being developed to use "lit" (or active) fibers as ocean observation platforms, greatly expanding the potential use of distributed acoustic sensing across the seafloor.

In addition, considering its preeminent role in developing and deploying advanced cyberinfrastructure, NSF's Directorate for Computer and Information Science and Engineering is an important partner for OCE in terms of cabled observatories and the Academic Research Fleet.

Biotechnology

The explosive growth of DNA sequencing, analysis, and manipulation, along with the keen interest in use of novel genes and gene pathways in creating new industrial-scale bioproducts, opens up the biotechnology and bioinformatics sector as a partner for future ocean exploration. For example, marine microalgae have been proposed as a crucial source of nutrients and food (Greene et al., 2022) for humans and for food production in aquaculture settings. Additionally, antifreeze genes from polar ice fish, which allow adaptation to subzero ocean temperatures, have been used across a wide range of industries (Eskandari et al., 2024). Furthermore, bioinformatics tools are widely used in collecting samples from ocean sediment and water microbial exploration. The speed of the advancement of genomic research, coupled with the rise of autonomous vehicles to collect samples, has led to the development of new autonomous DNA samplers.

Another area for partnership and technology transfer is the biomedical field. For example, *Prochlorococcus*, the most numerous photosynthetic organism on earth, was not discovered until flow cytometer technology, originally developed for medical applications, was used by oceanographers to analyze ocean water samples (Chisholm et al., 1988). DNA sequencing technology developed for the Human Genome Project fundamentally changed marine microbial oceanography by introducing shotgun metagenomic sequencing methods for ocean research (Rusch et al., 2007). And the reverse flow of ideas and methods has also occurred, with researchers trained in marine microbial ecology contributing to human microbiome science (Dubilier et al., 2015). The biomedical field, specifically partnership with the National Institutes of Health, offers additional opportunities for OCE to play a strategic role in enabling new resources and transfer of knowledge to the oceanographic community.

In all of these cases, basic discoveries in ocean sciences (e.g., heat-tolerant enzymes, new types of symbioses) are coupled with an industrial capacity to create products of value to society. Fostering these connections emphasizes the important role that ocean biotechnology plays in discovery and provides a way to augment the physical and chemical ocean sensing work done by other sectors within the field of ocean sciences.

Ocean Renewable Energy and Workforce Development

Developing new technologies and training a workforce on ocean-based renewable energy presents a ready opportunity for cross-directorate collaboration, for example through the NSF Clean Energy Initiative. Development of new instrumentation, maintenance of and improvement to equipment related to ocean-based systems, and data management efforts to ensure the quality and accuracy of the data and the analysis of large datasets generated by these systems all require a workforce with specialized training in ocean sciences, engineering, manufacturing, and data science.

Aquaculture and Mariculture

Several examples of NSF-funded projects to advance research and workforce development relate to aquaculture. The NSF Convergence Accelerator program has supported research on commercial marine fish production by developing sustainable feeds, establishing feed suppliers, and enhancing market acceptance. Another example relates to public policy, where NSF has supported research to examine the effectiveness of marine aquaculture partnerships measured by the emergence of new social capital, learning and consensus on scientific and policy issues, formal policy agreements, policy adoption by higher authorities, and projected socioeconomic and ecological outcomes. Third is the NSF TIP Directorate–funded research project BioSPACE: Biosensing Surveillance of Pathogens in Aquaculture and Coastal Environments. This project is focused on the development of a data analytics platform that uses biosensors to detect harmful organisms, such as Vibrio and Pseudomonas, in aquaculture farms and coastal waters. Aquaculture is driven by public demand for seafood protein, products and materials, and is therefore inherently opportune for partnering with the seafood and related industries. This field is also an important opportunity to engage subsistence fishing communities, or those with local and Indigenous knowledge on the species and their response to ecosystem changes.

The U.S. Merchant Marine

It has long been recognized that the ships of the U.S. Merchant Marine and other such organizations could be equipped with instrumentation requiring minimal human intervention to acquire routine oceanographic data in key locations (Rossby et al., 1995). For example, the *M.V. Oleander* has been operating between New Jersey and Bermuda for several decades, collecting oceanographic data. With NSF and other federal agency funding, the *Oleander* was equipped with an acoustic doppler current profiler and the capability to deploy expendable bathythermograph sensors. The result has been a valuable multiple-year time series of temperature and vertically resolved currents measurements across the Gulf Stream (Rossby et al., 2017)—one of the most important currents in the ocean. Additional partnerships with the Merchant Marine are an opportunity to build even more time series of important oceanographic variables, such as vertically resolved currents, temperature, salinity, and light detecting and ranging (LIDAR)-determined ocean particulate material. The data are most valuable when the route of an instrumented merchant ship is known far in advance, and the ship makes repeated crossings on the same route.

Additional Opportunities and Resources for Partnering with Industry

There is a pronounced need for targeted programs to strengthen the capacity of the academic ocean science workforce, in order to connect more productively with industry and local partners, and to engage effectively with innovation and entrepreneurship activities. Such programs can take the form of professional development activities or of incentive programs. These programs could be geared towards enabling the academic workforce to reimagine applications of its work in the context of innovation and entrepreneurship. Academic faculty are well-trained in conceiving projects that are driven by hypotheses and research questions; however, translation of their work into technological advances and/or commercializable products requires unique skills and partnerships, which can be encouraged through targeted funding calls, such as the Accelerate Research Translation program, aimed at the ocean science community.

CONCLUSION 3.5: Partnerships inherently include multiple sectors and are an important component to the transdisciplinary research approach to ocean and Earth science. Developing strategic partnerships will be both necessary and critical to accomplishing the ambitious and urgent research portfolio put forward in this report, specifically in observing the ocean (including platforms, data, and technology development) across ocean physics, chemistry, biology, and geology.

- *The NOPP is an example of an existing program that fosters interagency and industry support for basic curiosity- and discovery-driven ocean science research. Partnerships such as NOPP, while sometimes challenging to implement, can lead to highly impactful programs being established.*
- *Creative ways of thinking about new partnerships can result in cost-effective ways to access the ocean. For example, partnerships with groups such as the U.S. Merchant Marine and the U.S. Coast Guard have the potential to allow scientists to acquire and sustain key ocean measurements in critical parts of the ocean without utilizing ship time in the U.S. Academic Research Fleet.*
- *Opportunities exist to partner with related mission agencies (e.g., NOAA, National Aeronautics and Space Administration, National Institutes of Health, ONR, and BOEM), as well as with other NSF directorates (e.g., TIP, Computer and Information Science and Engineering, and Biological Sciences).*

RECOMMENDATION 3.3: To bolster and leverage urgent ocean science research over the next decade, National Science Foundation's (NSF's) Division of Ocean Sciences (OCE) should explore various ways of bringing new resources to this work, including expanding partnership with other NSF directorates, related mission agencies, and other organizations. OCE should seek greater interagency cooperation through federal policies and mechanisms for leveraging NSF-funded basic ocean science research with those mission agencies. Developing new partnerships in ocean sciences, as well as other disciplines, may increase support available for the development of new technologies, data curation, solutions-oriented research, and the basic research that will fuel these developments.

4

Facilities and Infrastructure for the Next Decade

This chapter assesses the infrastructure needed to achieve the ocean science research of the next decade. The evaluation of needs is parsed into four categories: major infrastructure (Academic Research Fleet [ARF], Ocean Observatories Initiative [OOI], and scientific ocean drilling), additional supporting infrastructure, underutilized or emerging infrastructure, and cyberinfrastructure. As long-term major investments in ocean science infrastructure supported by the National Science Foundation's (NSF's) Division of Ocean Sciences (OCE), the discussions of major infrastructure include a high-level program assessment as well as the role of that infrastructure in supporting the urgent ocean science research portfolio.

The conclusions and recommendations in this chapter focus specifically on supporting the three urgent ocean science questions outlined in Chapter 2:

- Ocean and Climate: How will the ocean's ability to absorb heat and carbon change?
- Ecosystem Resilience: How will marine ecosystems respond to changes in the Earth system?
- Extreme Events: How can the ability to forecast extreme events driven by ocean and seafloor processes be improved?

The conclusions and recommendations are not intended to reflect all the possible ways OCE-sponsored research will use OCE facilities and infrastructure. Furthermore, a full suite of conclusions related to scientific ocean drilling can be found in the committee's interim report (NASEM, 2024b), and it is important that the present report be read in the context of that interim report.

Table 4.1 provides an overview of the infrastructure needed to accomplish these three urgent ocean science research questions for the next decade, as justified in the text that follows.

MAJOR FACILITIES

OCE-sponsored research relies heavily on facilities and infrastructure funded by OCE and other agencies, with the most expensive NSF-funded programs being the U.S. Academic Research Fleet, the Ocean Observatories Initiative, and the Ocean Drilling Program. The annual cost of operating these three facilities accounts for about 50 percent of the OCE budget; the operating cost of the ARF is the most expensive (Figure 4.1). Note that for OOI and NSF-owned vessels, the costs shown in Figure 4.1 do not include construction, which were covered by a separate NSF funding account (Major Research Equipment and Facilities Construction) nor development costs that were covered by the OCE core budget prior to construction completion. Costs associated with operating the vessels in the ARF have risen significantly in recent years driven by fuel costs, salary increases required to keep experienced and competent vessel crews, and addressing technology needs associated with modernizing the fleet. As discussed below, modern ocean facilities are crucial for the research related to addressing the urgent research questions presented in Chapter 2.

Scientific ocean drilling was covered in detail in the 2025 Decadal Survey interim report (NASEM, 2024b). Expanding from that foundation, this final report evaluates scientific ocean drilling in the broader context of ocean sciences. The discussion provides background information and an assessment of infrastructure needs only in relation to the three science questions proposed in Chapter 2. The committee recognizes that science priorities beyond the three research questions may have different infrastructure needs.

Academic Research Fleet

Oceanographic research vessels are the fundamental platform for the field of ocean sciences to conduct research in the past, present, and for the foreseeable future. NSF-supported scientists and researchers

73

funded by other agencies and the private sector have been using the ARF for decades, demonstrating how the fleet has contributed to research and how the fleet will support research in future.

NSF and other federal agencies oversee the ARF through awards to each ship-operating institution and via the 53-year partnership with the University-National Oceanographic Laboratory System (UNOLS). UNOLS matches vessel use requests with available assets for oceanographic research by implementing a scheduling framework and by providing a forum for collecting recommendations and community priorities for replacing, modifying, or improving facilities. At the time of this report, UNOLS and NSF coordinate the scheduling of scientific expeditions aboard 17 vessels [1]located at 14 operating institutions.

The ARF consists of a fleet of oceanographic vessels (Table 4.2), three submersibles, and an autonomous vehicle, which are owned and operated by a combination of NSF, the Office of Naval Research (ONR), universities, and private entities. Vessel operations are supported through a contract with the agency that owns the vessel. The responsibilities of the contractor, generally an academic institution, includes crewing the vessel, providing a home port, and arranging the shipping of scientific gear and supplies to and from the vessel. All vessels in the ARF must be operated in accordance with UNOLS safety standards, undergo inspections, address performance standards, fulfill reporting requirements, and be accessible to the scientific community for the purpose of supporting national interests in marine science and engineering research and education and for fulfilling agency missions. Recently, cybersecurity has also been recognized as increasingly important and has become a requirement for ship operators. OCE recently created a new program officer position dedicated to improving cybersecurity for the ARF.

TABLE 4.1 Infrastructure Needs to Address Urgent Priorities for Ocean Sciences Research, 2025–2035

	Ocean and Climate	Ecosystem Resilience	Extreme Events
Major Facilities			
Academic Research Fleet (ARF)[a]	R	R	R
Scientific ocean drilling capability	R	G	R[b]
Ocean Observatories Initiative Arrays			
Pioneer	R	I	G[c]
Cabled	I	NDE	R[b,c,d]
Endurance	R[d]	I	G[c]
Open Ocean	R[d]	I	G[c]
Infrastructure Assets			
Non-UNOLS, non-ARF institution-owned research vessels	R[d]	R	R
Marine laboratories and stations	R[d]	R	R
Autonomous platforms	R	R and NDE	R[c] and NDE
Geophysical Instrumentation	NR	NR	R[b]
Instrumented cables	R[d]	G	G[b]
Ocean biotechnology	NDE	NDE	NDE
Novel sensors	R and NDE	R and NDE	R and NDE
Cyberinfrastructure	R and NDE	R and NDE	R and NDE

[a] Including the National Deep Submergence Facility
[b] For geological extreme events (e.g., earthquakes)
[c] For water column extreme events (e.g., severe storms, changes to Atlantic Meridional Overturning Circulation)
[d] For heat absorption
NOTES: G = good, if a byproduct of a primary driver; I = important but not required; NR = not required; NDE = needs development or more evaluation in relation to the needs of the urgent science questions; R = required; UNOLS = University-National Oceanographic Laboratory System.

---

[1] This sentence was changed after release of the report to correct the number of vessels in operation.

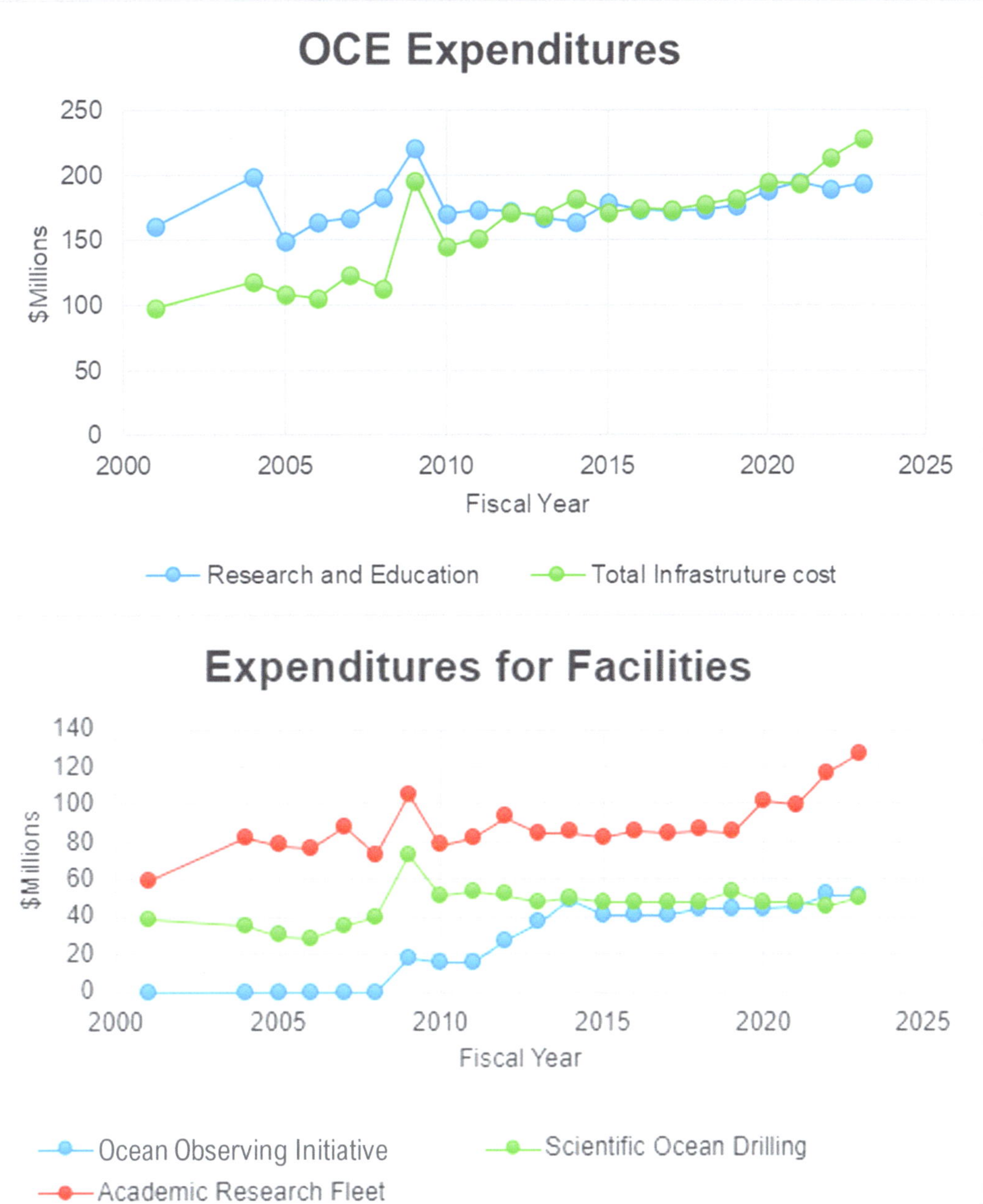

FIGURE 4.1 Time series of fiscal year actual expenditures for research and major facilities funded by the National Science Foundation's Division of Ocean Studies (OCE). NOTES: Upper figure 2023 data point represents appropriated funds.
SOURCE: NSF, n.d.-b.

The primary agencies funding the at-sea operations of the ARF are (in order): NSF, the Navy (ONR), and the National Oceanic and Atmospheric Administration (NOAA). From 2013 to 2023, agency-funded research on ARF vessels averaged a total of 3,000–3,500 annual days at sea, with each agency funding ship operating costs for the days their respective funded research programs used the vessels. Other operational funding comes from other federal, nonfederal, private, state, and other sources, including the operating institutions that annually purchase days at sea for educational and other uses.

TABLE 4-2 [2]Status of Ships in the Academic Research Fleet

Ship Class *Ship Name*	Year Built	Owner	Length overall m(ft)	Science Berths	Total Ship Days Used (2022)	End Year	Scientist- Days at Sea (est.)[a]	Host Institution
Global								
Thomas G. Thompson	1991	Navy	84 (274)	36	268	2036	9,648	University of Washington (UW)
Roger Revelle	1996	Navy	84 (274)	37	299	2041	11,063	Scripps Institution of Oceanography (SIO)
Atlantis	1997	Navy	84 (274)	37	244	2042	9,028	Woods Hole Oceanographic Institution
Marcus G. Langseth	1991	LDEO	71 (235)	35	216	2028	7,560	Columbia University
Sikuliaq	2014	NSF	80 (261)	26	264	2045[b]	6,864	University of Alaska Fairbanks
Ocean/Intermediate								
Kilo Moana	2002	Navy	57 (186)	29	259	2032[b]	7,517	University of Hawai'i
Endeavor	1976	NSF	56(185)	18	162	2026	2,916	University of Rhode Island
Atlantic Explorer	1982	BIOS	51 (168)	20	212	2026	4,240	Bermuda Institute of Ocean Sciences (BIOS)
Neil Armstrong	2015	Navy	73 (238)	24	276	2045[b]	5,424	Woods Hole Oceanographic Institution
Sally Ride	2015	Navy	73 (238)	25	173	2046[b]	4,325	SIO
Regional								
Hugh R. Sharp	2005	UDEL	44 (146)	14	163	2035[b]	2,282	University of Delaware (UDEL)
Taani[c]	2027	NSF	60 (199)	14	NA	2053[b]	NA	Oregon State University
Narragansett Dawn[c]	2027	NSF	60 (199)	14	NA	2054[b]	NA	University of Rhode Island
Gilbert R. Mason[c]	2028	NSF	60 (199)	14	NA	2054[b]	NA	The University of Southern Mississippi
Coastal/ Local								
Robert Gordon Sproul	1981	SIO	38 (125)	12	131	2030	1,572	SIO
Pelican	1985	LUMCON	36 (116)	14	220	2027	3,080	Louisiana Universities Marine Consortium (LUMCON)
Walton Smith	2000	Univ of Miami	30 (96)	16	50	2030	800	University of Miami
Savannah	2001	SkIO/UGA	28 (92)	19	135	2039	2,565	Skidaway Institute of Oceanography (SkIO)/University of Georgia (UGA)
Blue Heron	1985	UMINN	26 (86)	6	110	2030	660	Large Lakes Observatory
Rachel Carson	2003	UW	22 (72)	9	123	2033	1,107	Monterey Bay Aquarium Research Institute

[a] Estimated by multiplying "Science Berths" by "Total Ship Days Used"

[b] Lifetime expected to be extended by a mid-life refit.

[c] Regional Class Research Vessels are expected to be operational no earlier than 2027.

NOTES: NSF = National Science Foundation; LDEO = Lamont-Doherty Earth Observatory; UMINN = University of Minnesota.

SOURCE: https://UNOLS.org.

---

[2] This table was changed after release of the report to correct End Year data.

Current State of the Academic Research Fleet

From 2024 to 2030, the end of life is scheduled for three intermediate-class vessels (*Endeavor*, *Atlantic Explorer*, and *Kilo Moana*); four coastal-/local-class vessels (*Robert Gordon Sproul*, *Pelican*, *Walton Smith*, and *Blue Heron*); and a global-class vessel (*Marcus G. Langseth*). As an estimate of what capacity will be lost and what will be replaced with new vessels, the information in Table 4.1 (multiplying columns 5 and 6 and summing for the ships named above) suggests a potential loss of more than 32,000 scientist-days at sea per year. It is only an estimate, because this calculation assumes that the days at sea in column 5 of Table 4.1 are representative of future usage and that the ships sail with a full complement of scientists. The estimated losses will be partly mitigated when the new NSF-owned regional-class research vessels (RCR/Vs) *Taani*, *Narraganset Dawn*, and *Gilbert R Mason* (Figure 4.2) become operational, estimated to be no earlier than 2027. Assuming the new vessels will spend about 200 days at sea, they will provide about 8,500 scientist-days at sea. The committee notes some unique exceptions, that may, in fact, improve the gap in scientist-days at sea in the future. For example, with the *R/V Sproul*, scheduled to retire in 2025, Scripps Institution of Oceanography will be obtaining a new ship with funding by the state of California. The ship design has been completed, and construction will begin soon. Nonetheless, there remains a deficit from the upcoming retirements as described here.[3]

A loss of geographic distribution is also noted. In the Atlantic Ocean, *Endeavor*, *Atlantic Explorer*, *Smith*, and *Savannah* represent approximately 10,000 scientist-days at sea. *Narragansett Dawn*, when launched, will provide approximately 2,800 scientist-days at sea. *Blue Heron*, the only vessel in the Great Lakes, faces retirement without a replacement, which would leave no ARF vessel operating in the Great Lakes. In the Pacific Ocean, *Taani* will provide potentially 2,800 scientist-days at sea, while the retirement of *Sproul* and *Kilo Moana* will equate to a loss of more than 10,000 scientist-days at sea. The committee does not know of a replacement for the *Kilo Moana*. Following retirement of *Pelican* and *Walton Smith*, only one ARF ship, the new *RCR/V Gilbert R. Mason*, will operate in Gulf waters, supporting approximately 2,800 scientist-days at sea per year.

FIGURE 4.2 Three regional-class research vessels are under construction.
NOTES: Oregon State University is the prime contractor and will operate the first in the class (*R/V Taani*). The University of Rhode Island and a consortium led by University of Southern Mississippi/Louisiana Universities Marine Consortium will operate the second and third vessels in the class.
SOURCE: Oregon State University and Glosten Associates.

---

[3] The paragraph was updated after release of the report to correct some details about the new ship.

These losses impact the ability to conduct responsive research in the wake of coastal maritime disasters. It should also be noted that by the end of the decade, all local- and coastal-class ARF vessels will exit the fleet. Planning for new vessels is a long and costly process; based on recent ARF experience with the NSF-owned RCR/Vs, as well as the Navy-owned *Sally Ride* and *Neil Armstrong*, more than a full decade is required from "conception to implementation." Thus, these retirements pose an existential risk to achievement of the scientific goals outlined in this report.

> *CONCLUSION 4.1: In the coming decade, research vessel capacity will be reduced, owing to the end-of-life of multiple research vessels. The new NSF-built RCR/Vs will replace only some of the loss of scientist-days at sea from retired vessels, and their capabilities differ from many of the ships scheduled to be retired. Replacement plans of the Navy-owned* Kilo Moana *and various vessels owned by different state governments are not known, raising the possibility that ocean science research could be significantly hindered by a substantial decrease in the capabilities and assets available to scientists to conduct oceanographic research.*

Geographic Distribution in Vessel Operations

A vessel's home port is often a hidden driver on where and by whom scientific research occurs; it also determines who is exposed to shipboard experiences, starting as early as K–12 students. Broader impact efforts conducted as part of NSF-funded research programs and ship operation grants often speak to the ability to engage students and the public through experiential activities connected with the ships. The committee examined the geographic distribution of ship operations across the ARF to determine whether all scientists throughout the United States could participate in addressing the three research themes discussed in this report by utilizing a research vessel from the ARF in their home waters. Currently, most global- and ocean-class vessels are operated by institutions along the U.S. Atlantic and Pacific coasts (Figure 4.3), an appropriate distribution if they are only supporting the research along the respective coasts and farther offshore, as evidenced by ship utilization days (Table 4.2). This bicoastal utilization is set to continue based on current vessel replacement plans.

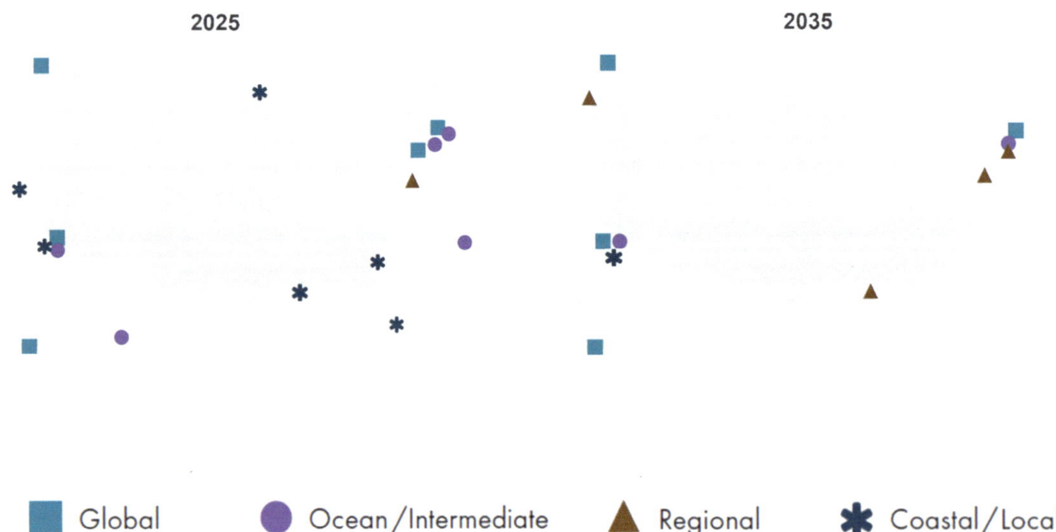

FIGURE 4.3 Aging vessel replacements within the Academic Research Fleet.
NOTES: Four global-class and two ocean-class vessels are owned by the U.S. Navy (*Kilo Moana, Atlantis, Thompson, Revelle, Armstrong,* and *Ride*). The end-of-life for *Atlantis, Thompson,* and *Revelle,* which underwent a midlife refit, will occur between 2036 and 2042. In addition, the two Navy-owned ocean class ships are due for a midlife refit in 2041 and 2042.
SOURCE: UNOLS Office.

However, ship-based research along the Gulf Coast, consisting of 1,680 miles (2,700 km) of coastline, is currently supported by only one coastal-/local-class ARF vessel and otherwise supported by a mosaic of aging coastal- and local-class vessels operated by academic institutions in the region. The largest and most capable non-UNOLS vessel operating is *R/V Point Sur*, a first-generation RCR/V retired from the ARF in 2015 and subsequently operated in the Gulf at The University of Southern Mississippi for the entirety of the last decade. This ship is scheduled for retirement upon the arrival of *R/V Gilbert R. Mason*. As evidence of the robust appetite for research in the Gulf, *R/V Point Sur*'s ship-based research days of operation in 2022 were 180, with 25 days supporting NSF-funded research. *R/V Pelican* and *R/V Point Sur* together have averaged over 400 days of operation (2020 and 2021 are exceptions, due to COVID) since 2015, including nearly 6,000 scientist-days at sea annually, displaying interest in oceanographic research in the U.S. Gulf Coast region.

NSF awarded a new RCR/V, the *R/V Gilbert R. Mason* (Figure 4.2) in 2019 to the region to be operated by the Gulf-Caribbean Oceanographic Consortium led by The University of Southern Mississippi and Louisiana Universities Marine Consortium (LUMCON). As the third build in this new class of ships, current project timelines place delivery and start of operations no earlier than the end of the current decade. The new RCR/V will provide an oceanographic ship capable of supporting research throughout the Gulf and surrounding open waters. *R/V Gilbert R. Mason* will then be the sole UNOLS vessel dedicated to supporting coastal and near-shore work in the U.S. Gulf Coast region, home to more than 200 million people.

By the end of the next decade, all ARF vessels located in the southeast United States other than the new *Gilbert R. Mason*, will have reached end-of-life. In addition, all other coastal-/ local-class vessels will have retired, including the vessel now operating in the Great Lakes. As pointed out above, location can determine what science occurs, where, by whom, and the communities that are engaged. This lack of available seagoing ocean science infrastructure may create scientific research and workforce inequities in the southeast United States, the Great Lakes region, and potentially the Pacific Islands due to a lack of NSF-supported opportunity in their respective backyards.

Another aspect of the geographic distribution in vessel operations is consideration of vessel capabilities in polar waters and adjacent seas. The NSF-owned and ice-capable global-class *R/V Sikuliaq* is an effective platform for studying ice-covered polar waters in both the northern and southern hemispheres. There is, however, a need for additional icebreaking capability to augment/replace the current aging U.S. icebreakers, particularly in the Antarctic region, as concluded by a recent NASEM report, *Future Directions for Southern Ocean and Antarctic Nearshore and Coastal Research* (NASEM, 2024d). The added capacity for enabling polar research, as described in *Future Directions* would support research in polar waters and help meet the challenge for the next decade and associated ocean science research priorities described in this report.

Oceanographic ships do not operate in isolation. Around vessel operations, other facilities (equipment and services) are needed to meet the needs of supporting shipboard science, thereby supporting a strong workforce and ocean economy in home port locations. Further, NSF currently supports 14 equipment and service pools or groups which, with few exceptions, also follow a bicoastal model, resulting in shipping costs and time delays for research in areas where these assets are lacking locally. NSF has highlighted geographic concentration of funding and resources in other directorates, with the goal of ensuring that research and innovation are not confined to geographical boundaries but are easily available across the nation (NSF, 2024a, 2024b). How geographical concentration of ocean science infrastructure impacts the pace and scope of ocean and coastal research may also be worth considering.

CONCLUSION 4.2: The committee recognizes the importance of access to infrastructure in order to support priority ocean science research. In addition to meeting the immediate needs of ocean research and education, over the long-term, the local and regional communities that host ocean research infrastructure benefit intellectually, economically, and socially from these national investments. For example, regions that support significant infrastructure are more likely to engage the citizenry with K–12 education, paired curriculum with nearby institutions of higher education, and develop a work-

force with high technical capabilities. Therefore, such investments offer an opportunity beyond science to strengthen multiple critical sectors that are of national interest and develop an ocean-literate citizenry that is well poised to engage in societal solutions.

Global-Class Research Vessels

Global-class research vessels are more heavily used than the other classes of vessels in the ARF (Table 4.2), meeting the critical need of research programs that require a large, interdisciplinary scientific party and/or a large platform for launching heavy equipment to study ocean processes in the global ocean. Global-class vessels can operate for longer periods at sea than the smaller classes of vessels and in more challenging sea state conditions. They also can carry a heavier load of deployable oceanographic instruments (e.g., current meter arrays) than other vessels, and *Atlantis* provides support for the NSF-supported deep submergence vehicles (*Alvin* and *Jason*). Multiplying the number of scientific berths by ship days at sea shows that the global-class ships support approximately 50 percent of the scientists at sea that use vessels scheduled by UNOLS (Table 4.2). The end-of-life of the three Navy-owned global-class ships (*Thompson, Atlantis*, and *Revelle*) occurs in the relatively short period of 2036–2042. Furthermore, the midlife refits of the Navy-owned ocean-class ships (*Armstrong* and *Ride*) occur only a few years later. Replacing these three global-class ships and refitting these two ocean-class ships is of critical importance to regaining U.S. leadership in the ocean sciences. Replacing the three global-class ships and refitting the two ocean-class ships within about a 6-year period will be a significant expense for the Navy. Now is the time to plan for this challenge, since a new oceanographic ship takes more than a decade from the start of serious planning until start of construction.

The global-class research vessel *Marcus G. Langseth*, owned and operated by Columbia University's Lamont-Doherty Earth Observatory (LDEO) with support from NSF, is the only ARF vessel capable of collecting the full range of active-source seismic data needed by the academic community. It also serves as the National Facility for Seismic Imaging. Currently, the *Marcus G. Langseth* is operational through 2024, with no announced long-term plan to replace it, although the committee was informed that discussions between LDEO and OCE are underway.

The need for global-class ships, while critical, is not new, and the fact that it persists speaks to the fundamental importance of seagoing research activities to fulfill national scientific objectives. Indeed, the National Research Council (2009) report *Science at Sea: Meeting Future Oceanographic Goals with a Robust Academic Research Fleet* stated, "The future academic research fleet requires investment in larger, more capable general purpose Global and Regional class ships to support multidisciplinary, multi-investigator research and advances in ocean technology" (p. 83). While NSF has fulfilled part of this with the new RCR/Vs coming online shortly, the portion of this recommendation regarding global-class vessels has only grown in importance and is now acute.

CONCLUSION 4.3: Three Navy-owned global-class ships (R/Vs Thompson, Revelle, *and* Atlantis*) and the* Langseth *face end-of-life in the near future. Midlife retrofits are scheduled for two Navy-owned ocean-class ships (R/Vs* Armstrong *and* Ride*). Refitting the ocean-class ships and replacing the four global-class vessels with those that have comparable or enhanced capabilities is necessary to accomplish high-priority ocean research over the coming decades and to regain U.S. leadership in providing researchers access to the ocean. The U.S. government needs to develop substantive planning and budget strategies now, as private and philanthropic seagoing opportunities are not considered true replacements for these vessels.*

Science Party Users

The range of operations and disciplinary breadth of users is a strength of the ARF, which hosts many different types of investigators on board, including atmospheric scientists, engineers, and informal science educators, which expands the pool of users among the science community. The science parties working on board ARF vessels typically include scientific faculty, staff, and students from various universities and research institutions, as well as skilled technicians to operate the sophisticated sampling devices and electronic systems required to collect data. Additionally, as Table 4.3 shows, students and educators (primarily K–12) are typically well represented in the makeup of the science party. Graduate students and postdocs are essential full-standing members of most science parties and are often deeply involved in research programs, doing much of the hands-on work at sea. NSF has made additional investments in providing support for early career training onboard vessels during select expeditions. These focused experiences help prepare early career scientists to write seagoing proposals and to be first-time chief scientists.

The U.S. scientific and ocean education community's access to time at sea via research vessels is invaluable to building a skilled ocean science workforce. Studies have shown that experiential learning (i.e., learning through hands-on experience) can lead to unique insights that are harder to achieve without firsthand experience. For example, a recent large-scale study by Lin and colleagues (2023) showed that remote collaboration fuses fewer breakthrough ideas than in-person, face-to-face collaboration. Lin and colleagues (2023) found that researchers working in the same space better integrate early career scholars into conceptual tasks, which can serve as a launching point for advancing scientific careers. In the ocean sciences, at-sea research programs create such spaces and thus need to be available broadly across geographies.

CONCLUSION 4.4: NSF has played an important role in ocean science workforce development by providing support for early career training onboard vessels. These focused experiences help prepare the next generation of ocean scientists and ocean science leaders. U.S. scientific access and the ocean education community's access to time at sea via research vessels is critical for building the future ocean science workforce.

Moreover, with the many advances in robotics and artificial intelligence, it is tempting to believe that lower-cost autonomous vehicles could do the work of a staffed research vessel. This is simply not the case for a number of well-established reasons. First, seagoing scientists are constantly adapting to unforeseen conditions and situations, and being at sea ensures that the data collected are most appropriate to meet their scientific goals. Second, many of the most profound discoveries in the field of ocean sciences have been made by scientists at sea. The committee posits that bringing the advances in autonomy, robotics, and computer science onto the research vessel will yield even more robust and thorough observations while still leading to ground-breaking and unexpected revelations. Third, currently and for the foreseeable long-term future, the physical capabilities of autonomous vehicles cannot fulfill the scientific mission of the three priority areas identified in this report because the capabilities of ships far exceed those of autonomous vehicles in many key areas. While there has been rapid progress in development of platforms, as noted elsewhere, sensors needed to make critical measurements are still lacking. Also, power, size, and telemetry challenges limit the number and kinds of sensors that can be deployed from an autonomous vehicle, and few can make the kinds of rate measurements needed for process studies.

TABLE 4.3 Participants Among Science Parties on Expeditions of the Academic Research Fleet

2012–2022	# of Participants
Educators/Outreach	502
Scientist, Postdoc	475
Student, Graduate	3,680
Student, Undergraduate	2,680
Student, K–12	32

NOTE: More than 1,000 institutions served by the Academic Research Fleet.

CONCLUSION 4.5: Currently and for the foreseeable future, while autonomous observations can augment shipboard measurements and perhaps reduce the length and cost of cruises, they cannot replace ship-based observational capabilities. Prioritization of developing new modes of sensing could reduce future reliance on shipboard measurements.

National Deep Submergence Facility

The National Deep Submergence Facility (NDSF) is a federally funded center based at the Woods Hole Oceanographic Institution, and the assets housed there are considered part of the ARF. The NDSF maintains and operates three deep-submergence vehicles: The human-occupied vehicle (HOV) *Alvin*, the remotely operated vehicle (ROV) *JASON II*, and the autonomous underwater vehicle (AUV) *Sentry*. These vehicles are state-of-the-art platforms that enable the U.S. scientific community to study deep-sea processes with the most advanced tools available. Each has unique capabilities. The *HOV Alvin* is a submersible that can take three persons, as well as a suite of user- and facility-owned sensors and samplers, on a half-day voyage down to 6,500 meters water depth. HOV dives provide the scientists with unequalled visual insights of the deep sea and the capacity to conduct sophisticated experiments that benefit from an untethered vehicle. The ROV *JASON II* is controlled via tether by a team of pilots and scientists from on board the ship that is operating the ROV. *JASON II's* power is provided through the tether so it can stay at depths down to 6,500 meters for days, allowing longer-term studies, experimentation, and extensive sampling. The AUV *Sentry* is an autonomous robot that can be programmed and deployed to sense and sample down to 6,000 meters. It is a highly capable vehicle used for autonomous mapping and imaging in the deep sea. It is also untethered and can be operated alongside the HOV *Alvin*.

The NDSF is an integral part of the U.S. oceanographic research infrastructure. It has evolved over the last decade, in terms of both its vehicles and operations. The HOV *Alvin* was rebuilt in 2014 and was recently upgraded by NSF to allow deployments to 6,500 meters depth. The ROV *JASON II* was modified several years ago to allow a heavy lifting capability and deployment configurations that make it amenable to being launched off both global- and ocean-class ARF vessels. The AUV *Sentry* was also recently upgraded to explore greater depths and to have longer-duration deployments and new modes of biological sampling, such as environmental DNA. The NDSF was also a pioneer in launching a "new user program" that facilitated the development of a larger user base. Additionally, the NDSF has worked to streamline the acquisition, curation, accessibility, and archiving of data (including video) collected by its assets to ensure that the data products are of the highest quality and broadly available. Recently, in response to community assessments, the NDSF has received support from NSF to build a smaller, more affordable midsize ROV to meet growing demand, especially in nearshore and coastal environs, which can be deployed from the new, NSF-owned RCR/Vs and other vessels.

Scientific Impact

The NDSF tracks scientific publications and other products associated with the use of its deep-submergence vehicles and has found that they have a marked impact on ocean science. These data are presented annually at town hall meetings. They reveal that the vehicles have played a key role in advancing ocean science in the last decade, including, but not limited to, the discovery of the deepest hydrothermal vents, discovering novel microbes in the deep subsurface biosphere, elucidating biogeochemical processes that support surface primary production, and unraveling the geophysical processes of plate motion. The NDSF has also been a launchpad for many careers, providing engineers and scientists with the opportunity to design and operate leading technologies. The Deep Submergence Science Committee sponsors workshops for early career scientists to acquaint them with the capabilities of the NDSF. Thus, the NDSF has historically contributed to workforce development, although there are opportunities for growth (presented below).

Current Funding Trends

The NDSF is supported primarily through a 5-year cooperative agreement with NSF (other support comes from NOAA, ONR, and other federal and nonfederal sources). Its funding is negotiated annually, based on the number of NSF-funded operating days. From 2014 to 2024, agency funding for the NDSF varied between $10 million and $13 million (depending on vehicle and research vessel refits, operational days, etc.). As is true for the ARF, operating costs have increased because of increases in salaries and cost of hardware and supplies. Moreover, all NDSF vehicles are deployed from research vessels, so increases in their costs have an impact on the overall day rate.

A number of issues have arisen over the last decade that have taxed the facility. In some years, the decreased number of vessels in the ARF, along with sustained demand for deep-submergence vehicles, has meant that some scientists wait 4 or more years to see their expedition realized. This is beyond the life cycle of a funding grant, burdening the scientist and NSF with a series of no-cost extensions, requests for supplemental support, and other administrative minutiae that contribute to inefficiency. There is also a persistent and growing interest in using ROVs and AUVs in the coastal and nearshore environment, but NDSF assets are often oversized for use in these locales and on regional vessels (e.g., the HOV *Alvin* can be deployed only by the R/V *Atlantis*). In addition, there are an increasing number of deep-submergence vehicles being operated by commercial, philanthropic, and other government entities. The directorate has, when appropriate, hired commercial operators and partnered with other nations to support global-scale research. U.S. scientists sometimes request these other operators to gain access to their study sites, especially where vessel support is limited geographically, and then turn to NSF to support the subsequent analyses. While this can be an effective model for an individual researcher or team, it is currently done ad hoc and, to the committee's knowledge, there is no standing policy about NSF's interest or ability to support such collaborations. Thus, NDSF's ongoing efforts to develop the midsize ROV need to be supported specifically.

CONCLUSION 4.6: The NDSF provides access to vehicles that are essential for research related to the report's three science themes listed in Conclusion 2.2. The following enhancements would improve the assets offered to the oceanographic community:

- *Expand the array of assets available to users, including lower cost ROVs and AUVs for deployments on coastal and regional vessels (i.e., used in nearshore and coastal regions).*
- *Expand the scientific footprint of each HOV, ROV, or AUV dive by enhancing remote science capabilities and the inclusion of autonomous assets that can work alongside the vehicles and perhaps be deployed simultaneously.*
- *Increase access to deep-sea research for early career professionals, by continuing to host workshops for early career scientists, among other activities.*
- *Enhance international cooperation to enable U.S. scientists to lead global-scale research efforts, including but not limited to supporting the use of commercial, philanthropic, and foreign national deep submergence assets when either it is more cost-effective or when U.S. assets are otherwise committed.*
- *Develop a formal trainee program that serves the needs of the facility and the greater needs of the maritime science, technology, engineering, and mathematics (STEM) workforce.*

Supporting a New Decade of Research

Answering all three urgent science questions from Chapter 2 will require extensive use of the ARF (including the NDSF) for process studies and, in some cases, to obtain the necessary ocean observations by deploying moored instruments, AUVs, and ROVs. For example, large shipboard science parties on an ocean- or global-class vessel are required for carbon flux studies, whether they are of the natural system or related to applying marine carbon dioxide removal (mCDR) technologies, because measurements are required of the dissolved carbon system components (carbon dioxide [CO_2], hydrogen carbonate, etc.), as well as the organic phases (both dissolved and particulate). Supporting measurements needed include nutrient pools and their update rates as well as the particulate sinking flux measured with sediment traps, isotopes, and newly developing novel approaches. NDSF vehicles are also important for seafloor studies and deep-ocean studies, including those related to geological extreme events and critical minerals research.

CONCLUSION 4.7: The ARF, including the assets in the NDSF, are the backbone of OCE-funded ocean science research programs and are required infrastructure for addressing the portfolio of urgent ocean research over the next decade. OCE has an important role to play in regaining U.S. leadership in ocean sciences by providing access to the sea with research vessels.

RECOMMENDATION 4.1: The National Science Foundation's (NSF's) Division of Ocean Sciences and appropriate partners should conduct an evaluation that details the funding allocation for the Academic Research Fleet in order to make informed distributions of the limited resources available over the next decade and beyond. The evaluation should consider metrics such as days of use funded by NSF; use of local vessels for NSF-funded research; geographic distribution of assets; and alternative mechanisms for providing researchers with access to the coasts (including salt marshes) and sea, such as through marine laboratories or new partnerships.

The long-term status of global-class research vessels is concerning; the committee urges interested parties to initiate and/or continue significant and intentional planning for replacements. Without such planning and subsequent implementation, accomplishing high-priority basic and solutions-oriented science will be compromised, and U.S. leadership in providing access to the ocean will be curtailed.

Ocean Observatories Initiative

The mission of OOI is to provide real-time data in remote places to better understand the ocean, its complexity, and how it is changing because of natural and anthropogenic processes. This mission aligns the three research priorities in this report. Concomitantly, however, there is concern within the broader oceanographic community whether OOI remains the right tool for meeting ocean science objectives in a landscape of limited resources and notable technological innovation. As with other facilities, the committee finds that the community of users and the facility itself has not always been effective at summarizing and communicating their successes nor building a strong sense of community.

The OOI's stated science objectives are ocean climate variability, ocean food webs, and biogeochemical cycles; coastal ocean dynamics and ecosystems; global and plate-scale geodynamics; turbulent mixing and biophysical interactions; and fluid-rock interactions and the subseafloor biosphere. The committee received briefings on progress towards the objectives from the operators but found it difficult to get a sense of how much the research communities in these fields are engaged with and use the OOI data.

One approach to assessing impact is to look at published papers; however, in curating publications, it was difficult to identify output specifically resulting from the OOI. The OOI website lists over 400 publications that are sorted by observing array rather than scientific discipline. Metrics obtained via searches of the Web of Science and Dimensions databases as reported to NSF found 275 peer-reviewed references that used OOI data or infrastructure from 2013 through 2024. Manuscripts related to geology (primarily using data from the cabled array) and physical oceanography (primarily using the open ocean arrays) were more common than those from the subfields of biological or chemical oceanography. These statistics refer to only NSF-supported research, whereas other U.S. federal agencies such as NOAA and ONR, as well as non-U.S. agencies, such as the UK's Natural Environment Research Council, also support research using OOI infrastructure and data, and even if specifically ingesting OOI data, do not always cite OOI as the data source.

The number of OOI-related NSF awards from the core science programs (as distinct from the cooperative award to operate the OOI itself) peaked in 2016 (22 awards) and declined in subsequent years (3 awards in 2024), whereas the number of publications citing NSF support has increased since 2013 to more than 70 publications in 2024 (Figure 4.4). The OOI Facilities Board suggests that this trend is related to the freely available OOI data regardless of supporting agency and the use of funding sources other than NSF to analyze these data.

Enumerating publications is only one measure of determining OOI research contributions. Specifically, numbers of resulting manuscripts do not necessarily correlate with scientific impact—for example, one key manuscript can have a disproportionate scientific impact. However, it is clear that publications using and mentioning OOI data are continuing to grow, and continued growth in use of the data may help justify the investment of resources and funds.

Education and outreach are other areas that can be used to measure impact. Since 2018, OOI has expanded its reach through social media, adding thousands of followers on LinkedIn, Instagram, X (Twitter), and Facebook. The OOI Facilities Board documented a large number of education and workforce development activities using OOI data in the United States and elsewhere at the high school, undergraduate, and other higher education levels. These are encouraging trends.

Nevertheless, it would be advantageous for the OOI operators and community to consider whether and how they are serving the needs of the broad interdisciplinary fields they seek to serve. NSF needs to consider a careful evaluation of the facility in the context of this report and OOI's stated science objectives, as described below.

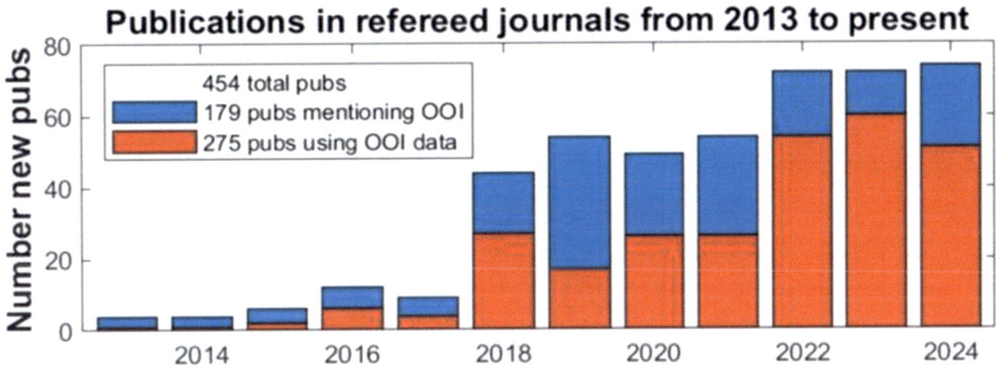

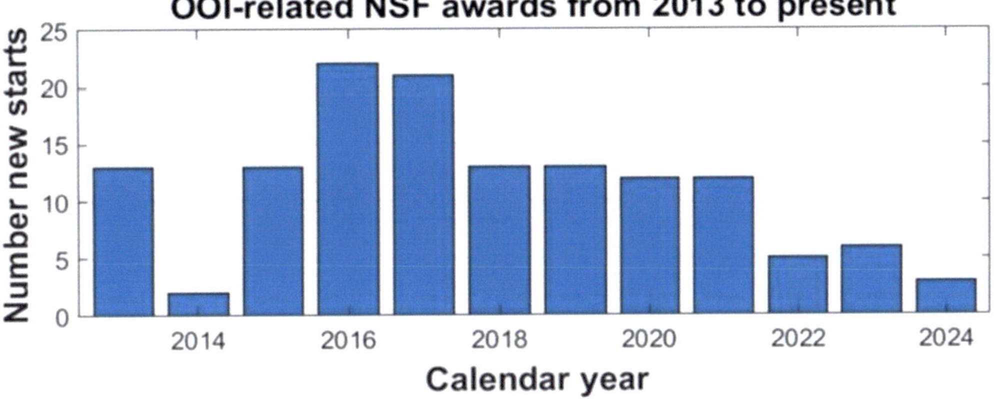

FIGURE 4.4 Publications citing data from the Ocean Observatories Initiative (OOI) (upper panel) and National Science Foundation awards for OOI research (lower panel).
SOURCE: OOI, 2025.

The committee's review of the OOI revealed a program that has many operational successes and strengths (Box 4.1) but also scientific gaps and a wider lack of coherence relative to the needs of the ocean science community. In the past decade, the OOI has worked out many of the programmatic aspects of the deployment and recovery of instruments and the support required to enable real-time services. These successes can be demonstrated in improved data collection and data reliability and some growth in publications arising from the instrument deployments.

The origins of what is now OOI began over 20 years ago, and the agency was initiated approximately a decade ago, when the needs for observations and the technological opportunities were much different than they are now. While the OOI's Implementation Organizations and the Program Management Office have kept up with technological innovation admirably, the overall fundamental premise of large, fixed assets as a research backbone may need to be reevaluated, given the changing state of ocean observations and framed in the context of the needs outlined in this report. Other approaches to ocean sampling (e.g., spatial sampling in extreme events such as hurricanes) also need to be considered, which could offer partnership opportunities to extend the footprint of observations and significantly enhance the adaptive sampling capacity. Such applications are being developed (e.g., using gliders to sample circulation patterns in the Pacific Ocean or in the path of hurricanes in the tropical Atlantic). Other complementary approaches should be considered for inclusion in the OOI, some of which are already underway, such as the use of SMART cables, distributed acoustic sensing, widespread ocean gliders, uncrewed surface vehicles, and more.

Some aspects of the OOI are very much aligned with the three priorities of this reports. For instance, the recently funded Cascadia Region Earthquake Science Center cable will monitor the unlocked portion of the Cascadia Subduction Zone, improving early warning and understanding of risk for extreme events; the long-term oceanographic data collected in the Irminger Sea is a critical site for understanding and forecasting changes in Atlantic meridional overturning circulation; the air–sea interaction measurements made with large open ocean buoys are relevant to the question regarding carbon and heat exchanges between the ocean and atmosphere; and coastal data provide vital context to important ecosystem and fisheries changes, from acidification to ocean heat waves to algal blooms. Nevertheless, a fixed system will never meet the extent of needs; consideration of the current OOI network in the broader context of nested observations— with careful thought to what is working, where gaps exist, and cost-benefit analysis—is needed.

CONCLUSION 4.8: Although the OOI program has enabled new discoveries, such as important seafloor and seismic processes around the Axial Seamount, and episodic events related to the Gulf Stream or the California Current system, there is a disconnect between the established program, the science achieved, and the current and future needs of the ocean science community broadly.

Supporting a New Decade of Research

The OOI, as it stands, would partially support the urgent research questions for the next decade. The cabled array component of OOI supports the research of extreme geological events, such as those that occur in the tectonically active zone off Oregon and Washington, and has the potential to collect additional oceanographic data, particularly using distributed acoustic sensing tools. There is every reason to believe that data from the array will continue to support research in the next decade. NDSF resources will be used to deploy and retrieve instrument packages to measure crustal strain and stretch. The deep ocean buoys in the Irminger Sea and at Station Papa, which focus on collecting data on air–sea exchange, climate, and the carbon cycle, provide insights into mechanisms governing heat and CO_2 uptake, and a better understanding of those processes, particularly under high wind regimes. When conducted near their locations, mooring and glider measurements at the Pioneer and Endurance arrays will provide key supporting measurements for process studies related to all three high-priority research questions. The Pioneer and Endurance arrays have also provided critical documentation of oceanographic extreme events related to ocean warming, changes in ocean circulation along the continental shelf in areas critical to commercial fisheries, and other processes operating over short timescales. Intentionally aligning future goals and capabilities of the OOI program with the needs of the ocean science community broadly and with the urgent science questions described in Chapter 2 would further increase the use of this major facility.

BOX 4.1
Ocean Observatories Initiative (OOI) Captures Submarine Eruption for the First Time

Axial Seamount hosts the most advanced submarine volcanic observatory in the ocean. A suite of 20 core OOI regional cable array instruments (Kelley et al., 2014) at the end of ~300 miles of cable links this most active submarine volcano off the Oregon–Washington coast to the internet (Kelley et al., 2015; Figure 4.5). In April 2015, only months after these instruments came online, Axial Seamount erupted just 4 years after its last eruption. Miraculously, the cable and instruments were undamaged, and the data collected provided multiple insights into volcanic eruptions and early warning signals that are applicable to both terrestrial and submarine volcanoes.

Increased seismicity, inflation, tremor, and tidal triggering preceded a seismic crisis associated with the eruption, which also led to the detection of an impulsive signal associated with lava erupting on the seafloor (Wilcock et al., 2016). Tilt and bottom pressure measurements detected deflation 1.5–2 hours after the start of the seismic crisis (Nooner and Chadwick, 2016), and approximately 1.5–2 hours after that, northward migration of some ~37,000 impulsive events signaled propagation of the dike along the north rift, ending with the eruption of a ~127-meter-thick lava flow (Kelley et al., 2015). Approximately a week after the start of the eruption, a series of diffusive broadband signals from 2 to 60 minutes in length were recorded and a significant increase in water temperatures near the floor of the caldera were recorded, coincident with an onset of diffusive activity (Caplan-Auerbach et al., 2017). Overall increases in bottom water temperature began 3 days after the start of the eruption and lasted for ~40 days (Xu et al., 2018). The increased temperatures are interpreted to reflect the emission of warm subsurface brines derived from the boiling hydrothermal systems at Axial (Xu et al., 2018). Monitoring of the caldera since the eruption shows direct linkages between changes in the inflation rate and seismic activity, and the volcano has inflated 90–95 percent since its last eruption (Chadwick et al., 2022). Current forecasts, updated near daily place the next eruption to occur between now and December 2025. Following lessons learned from the 2015 eruption, it is expected that a significant increase in seismic activity will predate this event, which has been observed since early 2024. The increased understanding of eruption precursors and signals is not only important for the Axial eruption; it has also led to the reinterpretation of a 2006 eruption along the East Pacific Rise (Tan et al., 2016), the timing of which had long been debated.

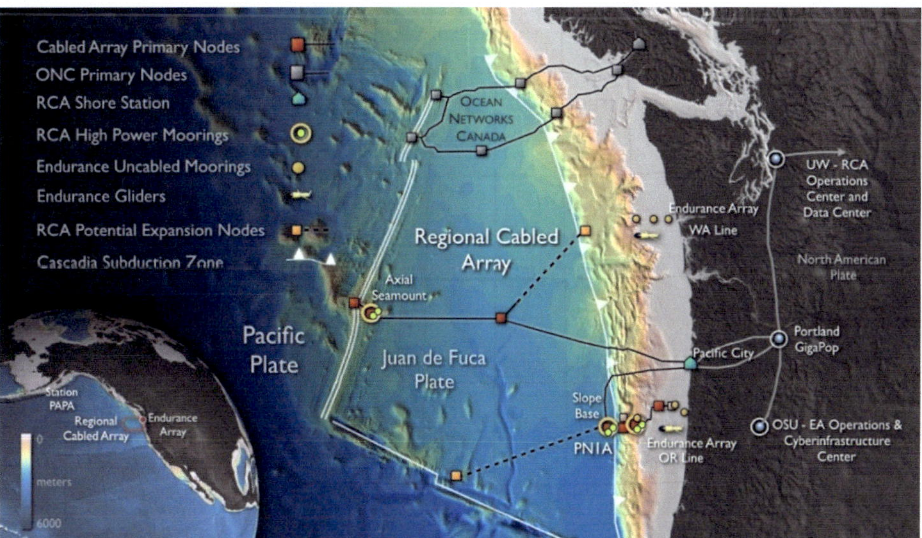

FIGURE 4.5 Schematic showing the cabled observatory, including the instrument locations on the Axial Seamount and the cable bringing real-time data to scientists located on shore.
NOTES: ONC = Ocean Networks Canada; OSU-EA = Oregan State University–Endurance Array; RCA = Regional Cabled Array; UW = University of Washington.
SOURCE: University of Washington.

RECOMMENDATION 4.2: The National Science Foundation's Division of Ocean Sciences (OCE) should conduct a revisioning and restructuring exercise for the future of the Ocean Observatories Initiative (OOI). The review, which should occur as a fully separate activity from the usual renewal/recompete discussions, could include:

- **an analysis of the scientific contributions of each of the OOI arrays;**
- **reconsideration of the goals and objectives of the program to better address the needs of the ocean science community and align with the evolving and urgent ocean science questions for the next decade; and**
- **consideration of how to incorporate technology that may not have existed when OOI was originally envisioned, including innovative ways to observe and measure biological abundances and processes, such as low-cost distributed observational networks.**

This recommended review differs from, and should be independent of, the renewal process that is focused on how well the infrastructure is meeting the requirements of the existing cooperative agreement. Rather, this investigative approach would provide a timely path forward for the next iteration of OOI that takes advantage of strategic partnerships, the evolution of ocean sciences and technological capabilities over the two decades since the OOI was originally conceived, and the approaches needed to address the urgent research priorities laid out in this report, as well as other science questions. This recommended review should occur well before the initiation of the review/renewal discussion regarding the current cooperative agreement.

Ocean Drilling Program

The United States has been the leader of international scientific ocean drilling for decades. During the second International Ocean Discovery Program (IODP-2) (2014–2024), 95 percent of all cores recovered were drilled by the U.S.-operated *JOIDES Resolution*, compared with only 4 percent with the European-contracted mission-specific platform approach, and only 1 percent with the Japanese drilling vessel, the *Chikyu* (see Table 1.1 in NASEM, 2024b). The United States was also the leader in expedition-related research publications, with more peer-reviewed journal articles between 2003 and 2021 than any other country (Figure 2.7 in NASEM, 2024b). Additionally, the United States was a global leader in achieving gender balance and in providing opportunities for students and early career researchers in ocean-going expeditions: since 2020, there were more U.S. women sailing on expeditions than men, with one-half of the expeditions led by women chief scientists; and there were equal proportions (one-third each) of graduate students, early career researchers, and senior researchers, on U.S.-operated expeditions during IODP-2 (Box 2.2 in NASEM, 2024b). These examples of leadership resulted from a uniquely capable infrastructure; a ready U.S. scientific workforce; and a culture of scientific excellence, mentoring, and collaboration (NASEM, 2024b).

As the majority funder of the *JOIDES Resolution*, providing $48 million per year through a cooperative agreement with Texas A&M University, NSF can take credit for much of this success. Rising costs, flatlined U.S. program funding, and a decrease in international partner contributions (from $16.5 million in fiscal year [FY] 2015 to approximately $5 million for FY2024; Figure 4.6) have contributed to a crisis in scientific ocean drilling that specifically and seriously threatens the future of U.S. leadership, U.S. scientific workforce development, and the advancement of ocean sciences overall. With the 2024 demobilization of the *JOIDES Resolution,* which was the only U.S. dedicated vessel for scientific ocean drilling and the only U.S. vessel capable of coring deeper than 50 meters into the seafloor, it is estimated that at least 90 percent of current scientific ocean drilling objectives identified in the 2025 Decadal Survey interim report will not be met (NASEM, 2024b). These unmet drilling objectives include those integral to address the three urgent priorities for ocean science research over the next decade: ocean and climate, ecosystem resilience, and extreme events (see Chapter 2 of this report).

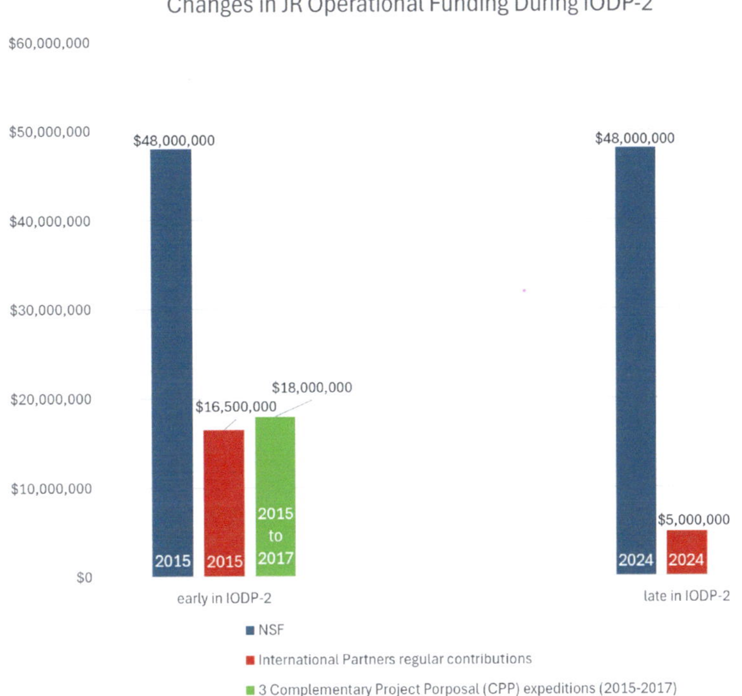

FIGURE 4.6 Changes in *JOIDES Resolution* operational funding streams during the second International Ocean Discovery Program (IODP-2). NOTE: JR = *JOIDES Resolution*; NSF = National Science Foundation. SOURCE: IODP, 2024, p. 7.

In order to successfully emerge from this crisis, a reenvisioned, sustainable U.S. scientific ocean drilling program needs to be defined, funded, and managed so that it can (1) maximize the scientific value of legacy assets (e.g., cores, data) already collected from scientific ocean drilling, (2) meet the need for long-term replacement of globally ranging deepwater drilling capabilities, and (3) encourage collaboration and fair international partnerships.

A pilot concept for multiyear, expedition-scale Legacy Asset Projects (LEAPs)—proposed by the IODP *JOIDES Resolution* Facility Board and currently managed by the IODP Science Support Office—was described in the interim report (NASEM, 2024b). A dedicated funding line at NSF for LEAPs would better incentivize the use of existing cores and create a funding structure for supporting it. Funding the LEAPs program will support basic ocean science research using existing cores, foster discovery and innovation, and create a mechanism to at least partially support and maintain a ready U.S. workforce for future scientific ocean drilling. However, without dedicated NSF funding, the LEAPs program will not facilitate future research as well as it could. The committee believes a funded LEAPs program, similar in funding level to the IODP-3 Scientific Projects Using Ocean Drilling Archives program ($1–2 million/year; Camoin and Eguchi, 2024) is a great approach to maximizing the scientific value of legacy assets and complementing new cores collected. A committee, including ocean drilling scientists, that makes sample allocation decisions would be an important part of a funded LEAPs program.

With respect to item (2), both short- and long-term strategies are needed. There is an immediate need to explore options for a sustainable drilling program involving a single U.S.-operated drillship, either newly constructed or leased. The outcomes of the current Subcommittee for a New Scientific Ocean Drilling Platform, formed by the NSF Advisory Committee for Geosciences, provide needed information on the estimated costs and the capabilities necessary to support the urgent research priorities in this report and in the 2024 interim report. A sustainable, long-term scientific ocean drilling program may benefit from reintroducing the option for complementary project proposals (CPPs) and nurturing/developing options for industry collaboration using the U.S.-operated drillship, with the goal of maintaining an additional, sustainable funding stream to supplement NSF funding for science operations. Extra funding generated from an industry-based expedition in 2012, and from CPP contributions from interested partner countries in early IODP-2 (Figure 4.6), may be models to expand in the future. On the other hand, international participation in the original Ocean Drilling Program (which ended in the early 2000s), including partial funding of *JOIDES Resolution* operations, worked more successfully than the later various IODP-named programs; key aspects of that earlier model could be reconsidered. Additionally, as outlined in the interim report (NASEM, 2024b), different management and operational approaches may be needed to balance financial constraints with needed shipboard versus shore-based measurements, staffing to support operations (prior to, during, and after expeditions), and advisory structures for safety and scientific purposes.

In the short term, until a dedicated drillship is available, NSF needs to continue supporting the use of mission-specific platforms for addressing high-priority, urgent science questions. As identified in the interim report, the *JOIDES Resolution*, now no longer under long-term NSF contract, could be considered a mission-specific platform, possibly for an extended period involving multiple expeditions over several years. Nevertheless, the committee sees U.S. and international mission-specific platforms as a "bridge solution" because one-off mission-specific platforms are logistically challenging in terms of securing vessels, engineering, and technical expertise, and are thus expensive and do not alone meet the identified science objectives (NASEM, 2024b).

Finally, collaboration among scientists from different countries, disciplines, and career stages has been a hallmark of scientific ocean drilling for decades. Collaborative science is better science, and this principle remains true as the committee seeks a blueprint for the future of scientific ocean drilling. As NSF and the U.S. and international scientific community work to emerge from the current crisis, this principle needs to remain an anchor in deliberations on the next steps.

> *CONCLUSION 4.9: The interim report (*Progress and Priorities in Ocean Drilling: In Search of Earth's Past and Present*) includes many conclusions related to the future of scientific ocean drilling. Following the release of this report, OCE took several important steps towards planning for a future dedicated vessel with globally ranging scientific ocean drilling capabilities, as well as steps towards securing future mission-specific platforms for the ocean drilling community to use in the nearer term. OCE also committed to continue preserving and curating ocean drilling legacy assets (cores, samples, and data). Retaining the ability to support (financially and otherwise) U.S. scientists pursuing scientific research using new and archived ocean drilling assets remains critical to basic research and to the urgent ocean science research portfolio identified this report. Nonetheless, because of the decommissioning of the* JOIDES Resolution, *there will be substantive and significant unmet drilling objectives that are integral to addressing the three high-priority research questions identified in this report.*

Supporting a New Decade of Research

The 2025 Decadal Survey interim report on scientific ocean drilling, *Progress and Priorities in Ocean Drilling: In Search of Earth's Past and Future* (NASEM, 2024b), identified three vital and urgent themes related to the three science questions discussed in Chapter 2:

1. Ground-truthing climate change: Advancing understanding of climate and ocean change drivers, feedbacks, and past tipping points.
2. Evaluating past marine ecosystem responses to climate and ocean change: Fossils can be used to determine ecosystem responses to ancient environmental changes, including warming, ocean acidification, and deoxygenation. This understanding of the past can be used to better understand the way that current, rapid environmental changes over decades generate ecosystem change.
3. Monitoring and assessing geohazards: Providing data to more accurately forecast and assess future risks of earthquakes, volcanic eruptions, submarine landslides, and tsunamis.

These three urgent and vital priority areas for a future scientific ocean drilling program align with, and are integral to, research over the next decade that is needed to address the three urgent ocean science questions identified by this committee. Additionally, while not described as urgent in the interim report, several other vital areas of research were listed. For example, scientific ocean drilling of the ocean crust (basalts) has yielded frontier research into the nature of the deep microbial biosphere and has provided foundational evidence on the evolution of the oceanic (and continental) crust and its role in the long-term geochemical cycling of elements. Research on the subseafloor biosphere has direct implications for understanding the potential for life in other areas of the solar system, the origins of life on Earth, and the integral building blocks of ecosystems that nurture the biological world. This research is also necessary for identifying potential environmental effects from the harvesting of critical minerals from the seafloor. The cycling of fluids through the subseafloor and corresponding chemical exchanges between the liquid and solid earth may impact processes with direct societal relevance, including the production of mineral resources, sequestration of atmospheric CO_2, and origin of geohazards (including volcanic eruptions, earthquakes, and related tsunamis). These are discussed in greater detail in the interim report (NASEM, 2024b).

CONCLUSION 4.10: Research utilizing scientific ocean drilling plays a unique and important role in addressing the urgent science priorities:

- *Ocean and Climate: How will the ocean's ability to absorb heat and carbon change?*
- *Ecosystem Resilience: How will marine ecosystems respond to changes in the Earth system?*
- *Extreme Events: How can the ability to forecast extreme events driven by ocean and seafloor processes be improved?*

The assets that will be available for scientific ocean drilling for the next decade are not yet known. In the past, by deploying sophisticated instrument packages in boreholes and in other ways, scientific ocean drilling has made significant contributions to understanding earthquakes and other extreme seafloor events. Furthermore, analysis of cores collected throughout the global ocean have documented changes in ocean carbon cycling, times of rapid warming and extreme heat, and identified periods of significant ecosystem changes related to the biology, chemistry, and physics of the ocean, such as past periods of extensive deoxygenation and ocean acidification associated with ocean and atmospheric warming. These studies provide insights into the impacts of a future warming planet. This research may continue in the future using the extensive archive of cores previously collected but also depends on new cores from future drilling operations. Without new approaches to addressing both legacy assets and infrastructure needs, the capacity for future U.S. operational and scientific leadership is bleak. It is unlikely that such U.S. leadership can be regained without a U.S.-based drillship.

RECOMMENDATION 4.3: The National Science Foundation (NSF) should take action to regain U.S. leadership in scientific ocean drilling on a global stage. To support basic ocean science and the urgent ocean science research portfolio identified in this report for the next decade and beyond, the committee makes the following recommendations:

- **Legacy assets: NSF's Division of Ocean Sciences (OCE) should create a dedicated funding line for the Legacy Asset Projects program to support expedition-scale collaborations, which maximize the return on legacy assets by providing funding to scientists to conduct large-scale research with existing cores and data.**
- **New drilling infrastructure:**
  - **OCE should develop a sustainable ocean drilling program, taking into account the need for a U.S.-based drillship. The committee urges NSF to creatively implement management and operational models that ensure viable long-term operations of such a vessel, including developing and nurturing more options for industry work. Such new management and operational models may differ significantly from the recently concluded drilling program.**

    o   **Until a dedicated drillship is available, OCE should continue to support the use of mission-specific platforms for addressing high-priority, urgent science questions, and to the extent possible, support efforts to retain the technical and engineering expertise in deep-sea coring previously employed by the United States with the *JOIDES Resolution*.**

- **International collaboration and coordination:**
  - o   **NSF and international funding agencies and governments should coordinate and collaborate globally toward an integrated, long-term strategy for scientific ocean drilling. Such collaboration will require meaningful and transparent reciprocity in scientific participation levels, financial support, scientific planning, and more among contributing partners.**
  - o   **OCE should renew dialog with both new and long-standing international partners to identify cost-sharing models for dedicated drillship operations; such models may include scientist shipboard participation proportional to international contribution levels.**

SUPPORTING INFRASTRUCTURE

Oceanographic Institutions and Local Research Vessels

Oceanographic institutions support NSF-funded research in many ways, including by supplying research vessels and access to marine laboratories and stations. For example, the ARF is not the only mechanism for providing ship time, and the vessel usage described in Table 4.2 is an incomplete picture; however, the committee did not have information as to how much other NSF-funded research is conducted on board vessels that are not part of the UNOLS-coordinated ARF. Tracking the amount of OCE-funded research taking place on smaller vessels, such as those operated by the Virginia Institute of Marine Science (College of William and Mary), The University of Southern Mississippi, and Florida State University, would clarify the contributions of these vessels to the OCE research portfolio. This analysis would be timely, as all the local- and coastal-class vessels in the ARF will reach end-of-life by the end of the decade. This suggests NSF-sponsored research in estuaries, bays, and coastal waters may have to rely more heavily on local vessels that are not currently in the ARF.

The committee attempted to conduct an informal review of institutions operating outside of UNOLS to provide days at sea in coastal waters on the Pacific, Atlantic, and Gulf Coast. The aim of this effort was to obtain an understanding of how much NSF-supported science was occurring on board these vessels. The response rate for this inquiry was low, however, probably hampered by a lack of knowledge by the vessel operators of the funding sources of those that come on board to conduct research. Some institutions reported the affiliation of scientists on board or the research activities that took place, but information on how NSF was using these ships to support funded research could not be derived from their responses.

CONCLUSION 4.11: Evaluation metrics, such as days of use funded by NSF on non-ARF local vessels reported by the ARF community, would be helpful to OCE for making informed allocations of the resources available.

Marine Laboratories and Stations

U.S. marine laboratories and stations constitute a large, diverse, and critical element in U.S. ocean science, teaching, and infrastructure investment. As discussed in Chapter 3, 99 labs are members of the National Association of Marine Labs, situated in 36 states and territories. The labs have access to unique and important marine research areas in the Pacific, Gulf, Caribbean, Atlantic, and Great Lakes (Figure 4.7). NSF supports a specific Field Stations and Marine Laboratories infrastructure improvement program, and about half of marine labs received funding from this program during the last 15 years.

FIGURE 4.7 Locations of the 99 labs that are members of the National Association of Marine Laboratories. NOTES: Most labs are in coastal North America, with some in the tropical Pacific and Caribbean. SOURCE: National Association of Marine Laboratories.

In addition to the vessels described above, marine laboratories and stations contribute to the infrastructure supporting NSF research by:

- providing small boats for research related to coral reefs, coastal wetlands, and estuaries;
- providing the space and flowing seawater systems to support experiments requiring tanks and aquaria for research projects;
- purchasing ship time on the ARF fleet to support research and education activities;
- supporting research library buildings, collections, and personnel;
- absorbing the construction costs of docks and piers;
- providing personnel, as well as vehicles and other equipment, to support loading and unloading operations;
- cost-sharing with NSF for large and expensive scientific instruments, such as mass spectrometers;
- providing space to store rock and core samples, and benthic and pelagic organism collections;
- providing storage space for field gear, as well as shipboard equipment, such as winches, and;
- collecting, curating, and storing coastal data on a long-term basis.

Research, training, outreach, and community connections at the U.S. marine stations are vital to support the three research priorities outlined in Chapter 2. First, integration of physics, chemistry, and biology is a key feature of the report, and this mix is relevant to the climate change work going on in coastal zones, particularly at U.S. marine labs. Second, of central interest to marine labs is integrating physical and chemical models into ecosystem processes, a line of research that is highly relevant in the coastal ecosystems of bays, estuaries, marshes, and sea grasses that are nearby many marine stations. Third, long-term ecosystem and ocean stability is key to understanding extreme events, and studying these is particularly cost-effective in coastal areas. As a result, the constellation of marine stations represents on-the-ground facilities that help stage ocean and ocean-life research across the country.

As described in Chapter 3 (Box 3.5), marine stations are also well suited to support development of the broad, transdisciplinary research that was the focus of Chapter 2. The stations themselves are often part of a local community and can help integrate local, Indigenous, and traditional knowledge in research and education. In these ways, the marine stations are major conduits to progress in components of the 2025 Decadal Survey.

CONCLUSION 4.12: Marine laboratories, marine stations, and oceanographic institutions make significant contributions to the infrastructure and workforce supporting OCE-funded research; financial and administrative attention is needed for their continued support.

Autonomous Platforms

Recent years have been a watershed for the development of AUVs, autonomous surface vehicles (ASVs), and gliders (ultra-low-power AUVs). In the past decade, legislation supported collaborative development of these technologies among federal agencies (i.e., NOAA) and industry (Commercial Engagement Through Ocean Technology Act of 2018) to advance the pace of autonomous operational capabilities. This is a strong indication of support for the widespread development and potential use cases for these technologies. A quick search of Google Scholar[4] reveals that in 2014, there were 2,380 publications on ocean AUVs. By 2024, the number of publications had more than doubled to 6,310. Publications with ocean gliders rose by over 40 percent in that time frame. The U.S. ocean sciences community has been at the leading edge of developing these marine autonomous technologies, and they are in use at a wide array of institutions (although the usual disparities still exist, with better-funded institutions having greater access to these assets). Furthermore, the proliferation of low earth orbit satellites such as Starlink can vastly improve communication and telemetry from AUVs.

Measurements from autonomous vehicles and floats will be critical for research related to the three high-priority science questions (Table 4.1). Research using autonomous platforms falls within two general categories: (1) sustained observations using vehicles or floats, such as gliders; and (2) vehicles and floats to extend the profile of oceanographic ships for studies of ocean processes. Both types of applications are essential to future ocean sciences research, and use of the latter will likely grow in popularity in the future.

Sustained Observations Using Vehicles or Floats

The Argo program provides global measurements of heat content in the upper 2,000 meters (Box 4.2), and new Argo floats will have the capability to make those measurements in waters deeper than 2,000 meters (Roemmich et al., 2019). At present, the Argo program is critical for collecting data to determine how much heat the ocean absorbs and how the accumulated heat spreads across ocean basins at the global scale, and for providing data to support forecasts of future trends in ocean heat content. Although the Argo measurements related to biological processes (referred to as biogeochemical [BGC]-Argo) are not fully mature, they indicate phytoplankton biomass, particulate organic carbon, and attenuation with depth of solar radiation. An NSF-funded Science and Technology Center has provided multiple BGC-Argo floats for deployment in the Southern Ocean. These are important supporting measurements for studies related to ecological resilience and also for providing background variability of particulate organic carbon essential for validating and mCDR approaches. To the committee's knowledge, funding for BGC-Argo expires in 2025, and the future of this program is unknown.

CONCLUSION 4.13: The Argo program serves as the main source for collecting critical observations of the ocean's interior. OCE's continued role in the national and international partnerships that support and enhance the Argo floats, particularly BGC-Argo, will be essential to addressing the urgent ocean science research questions of the next decade.

---

[4] See https://scholar.google.com.

BOX 4.2
The Argo Program

The Argo program celebrated its 25th anniversary in 2024; this program provides sustained observations based on a network of autonomous profiling floats—more than 3,900 are deployed as of 2024 (Figure 4.8). The original floats profiled the upper 2,000 meters measuring temperature and salinity at 1–2 meters resolution during a 10-day cycle. Next-generation floats can now profile into deeper waters, and the biogeochemical (BGC)-Argo floats also measure pH, oxygen, nitrate, chlorophyll (from fluorescence), particulate organic carbon (from particle backscatter), and water clarity (from solar irradiance change with depth) in addition to temperature and salinity. Additionally, some floats have been instrumented with passive acoustic listening sensors, which can be used to estimate rainfall rate and wind speed (Yang et al., 2015). New floats with particle imaging sensors are being deployed to identify and quantify zooplankton. During the period of the 10-day cycle when the Argo floats are at the surface, data are transmitted via the Iridium satellite network to data centers.

Argo has provided more than 2 million temperature and salinity profiles in the global ocean, far exceeding the number of such profiles available prior to Argo program. More than 6,000 publications used Argo measurements to:

1. determine ocean excess heat uptake,
2. determine the warming effect to sea level rise,
3. improve calculations of ocean carbon dioxide (CO_2) uptake with seasonal resolution,
4. determine effect of storms for enhanced CO_2 uptake,
5. improve understanding of global ocean acidification, and
6. determine monthly changes in global oxygen and particulate carbon as indicators of ecosystem productivity.

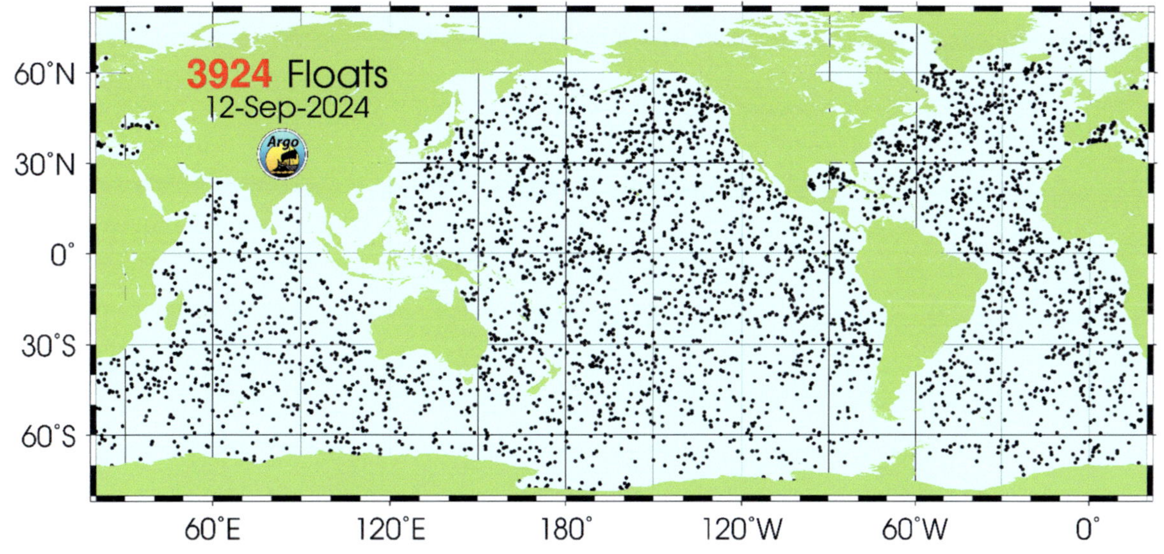

FIGURE 4.8 Global distribution of Argo floats as of September 12, 2024.
SOURCE: https://argo.ucsd.edu/about/status.

Gliders, either operating as part of individual grants or as part of larger initiatives (e.g., OOI), provide measurements comparable to those from Argo floats, including temperature and salinity and other suites of sensors, such as biogeochemical or acoustic arrays. Their ability to support adaptive sampling plans makes them particularly useful for characterizing the environment surrounding process study locations, for all three of the priority science questions identified in this report.

Another mechanism for obtaining sustained autonomous measurements, which has potential for future public–private partnerships, is the "fee-for-service" model. This observational model is different from other autonomous platforms in that it provides data without involving the sale of physical hardware. It is not clear how principal investigators are treating data collected by fee-for-service companies using their NSF funds;

risks include uncertain data availability and transparency of the data generation process. This model has been used in missions to improve weather prediction and observe extreme weather events in harsh conditions such as in the Southern Ocean and inside major hurricanes (Zhang et al., 2023) (see Figure 4.9). Companies offering the fee-for-service model may be an avenue for proliferating the use of autonomous platforms for sustained ocean observations.

Autonomous vehicles can significantly reduce the burden on the ARF by providing acoustic GPS. Traditionally, acoustic GPS beacons on the seafloor must be surveyed by ship once or twice a year to determine an accurate position. However, new approaches are being developed that use ASVs for surveying instead, freeing up oceanographic vessels and allowing continuous geodetic monitoring of the seafloor.

Vehicles and Floats Can Extend the Profile of Oceanographic Ships

AUVs and floats can expand the footprint of oceanographic ships, which is particularly important when the research focuses on rapidly evolving features (compared with the maximum speed of a ship), when they are 10s to 100s of kilometers in scale, and/or when adaptative sampling is required. A single ship can either do detailed measurements at a single location, stay at a single location to observe change with time, or sample along a track or grid, but cannot do all simultaneously. Deploying AUVs and other devices not tethered to the ship during the research expedition can help accurately determine changes in both time and space of ocean features and be particularly valuable to geographically focused process studies important to progressing the urgent science research portfolio for the next decade. These platforms are expensive and require specialized expertise to operate, which takes up valuable scientific berthing space. One potential way to minimize the cost while maintaining capability is to obtain a fleet or pool of AUVs that can be assigned to different expeditions and/or partner with entities—such as the Naval Oceanographic Office (NAVO) and NOAA's Uncrewed Systems Division under the Office of Marine and Aviation Operations—to share resources, training opportunities, and crew. While many institutions are well known for their respective fleets of autonomous vehicles that contribute to ocean observing, the NAVO also operates a fleet of autonomous vehicles, including gliders, which have been instrumental in improving Atlantic hurricane forecasts (Zhang et al., 2023).

FIGURE 4.9 Saildrone that recorded a 196-day time series of the flux of carbon dioxide at the surface of the Southern Ocean.
SOURCE: Image courtesy of Saildrone, Inc.

CONCLUSION 4.14: Ocean science increasingly relies on autonomous platforms to make important oceanographic measurements. Since these platforms will continue to play an important role in answering the urgent research questions of the next decade, consideration for improving their accessibility is needed. For example, additional assets could be secured for use by NSF-funded project teams by forging new partnerships, or through the creation of a national glider and small AUV pool. In addition to securing new assets, there are increasing opportunities to partner with commercial entities that sell oceanographic data rather than observations platforms.

Geophysical Instrumentation

The original Ocean Bottom Seismograph Instrument Pool was established with OCE funding in 1999 to provide broad community access to a new fleet of short-period and broadband ocean bottom seismographs; this program was later reestablished as the Ocean Bottom Seismic Instrument Center (OBSIC) in 2018. OBSIC currently operates a fleet of 106 deployable ocean bottom seismographs, of which 25 are short period and the remaining 81 are broadband units of varied type. In mid-2025, the construction of another 35 broadband ocean bottom seismographs, funded with a 2021 NSF Mid-scale Research Infrastructure award will be completed. A critical future need is increasing the short-period instrument pool, with additional instruments that can be deployed on the seafloor for durations of 2 months to 1.5 years. This pool of ocean bottom seismographs is in sustained demand from the scientific community. Their data have enabled detailed structural images of midocean ridges, hotspots, subduction zones, oceanic and continental transform faults, oceanic lithosphere and upper mantle, subsurface magmatic systems, and gas hydrate deposits. In addition, ocean bottom seismograph data have been used for monitoring seismicity and characterizing rupture properties of earthquakes at midocean ridge spreading centers, transform faults, and subduction zones. Over the last decade, earthquake detection and location algorithms have improved considerably, leveraging new methodologies using machine learning and artificial intelligence. Marine seismic data also provide critical imaging for structural and stratigraphic characterization in conjunction with scientific ocean drilling. This synergy led to important advances in subduction zones geohazard research, including an improved understanding of plate coupling related to slow slip and megathrust earthquake processes, and the origin of submarine landslides.

To resolve tectonic motions on the seafloor, NSF's Mid-scale Research Infrastructure program established the Seafloor Geodetic Instrument Pool (SGIP). Seafloor geodetic instrument–equipped acoustic beacons are an emerging technology with great future potential for widespread use, and the data provided are critical for assessing seismic hazards associated with subduction zone megathrust faults. In addition, absolute pressure gauges can be co-located with the acoustic GPS beacons to determine vertical positions, although instrument data drift limits the ability of these sensors to resolve signals longer than 1 month (e.g., fault creep or magma chamber inflation). Based on current technology, battery life limits deployments to about 10 years. Given the small number of instruments available and the need for long-term deployments to resolve tectonic signals, the Seafloor Geodesy community held two workshops to determine the location of an initial community experiment that would utilize the SGIP (Newman et al., 2021, 2022). The outcome of these workshops was an experiment offshore of the Alaska and Cascadia margin that will run for an initial period of 5 years using two-thirds of the SGIP transponders. At the time of this report, unfortunately, no funding mechanism is available to replenish the instrument pool following these deployments.

CONCLUSION 4.15: Seafloor geodetic instruments and ocean bottom seismometers provide data essential to understanding extreme events originating on the seafloor, such as earthquakes, subsurface landslides, and subsurface volcanism. The number of seafloor geodetic sensors and short-period ocean bottom seismometers are currently insufficient to meet community needs. Additionally, the R/V Marcus Langseth is nearing end-of-life in 2026, at which point no vessel will be capable of collecting active seismic measurements.

UNDERUTILIZED AND/OR EMERGING OBSERVATIONAL TECHNOLOGIES

Acoustics

Acoustics is an excellent and often underutilized methodology for ocean science and has the potential to emerge as a cost-effective and mainstream approach for collecting persistent ocean observations key to answering many of the urgent science research questions of the next decade. Although the use of acoustics for underwater sensing has been known since the time of Leonardo DaVinci, this methodology has been used primarily for national security. However, recently there has been a resurgence of interest in the application of acoustics for environmental measurements and monitoring (NASEM, 2024a). For example, since the speed of sound depends on temperature, salinity, and pressure, acoustic experiments can determine changes in ocean processes such as in the Arctic (Worcester et al., 2020). In addition to the traditional method of measuring ocean temperature through propagation, which can be done on a global scale (Dushaw et al., 2009), new techniques are being developed to use seismic activity to measure ocean temperature without a dedicated source (Wu et al., 2020). Furthermore, wind-driven ambient noise can be used for such disparate applications as monitoring rainfall in the open ocean using specially instrumented Argo floats (Yang et al., 2015) or for probing the seabed as a passive fathometer (Siderius and Gebbie, 2019). For decades, acoustic measurements have been used to detect seismic activity for early warning systems, but recent advances, such as instrumenting existing telecommunication cables (Howe et al., 2019) and distributed acoustic sensing, can amplify this capability. These technologies, which can also be used for thermometry as described above, provide continuous data feeds. Sonar, such as multibeam echosounder, is the primary method of mapping the ocean floor, contributing to a global initiative to provide high-resolution bathymetry for the entire ocean floor.[5] There is large untapped potential for acoustics to observe ocean biologic processes in applications such as seagrass meadow monitoring and tracking marine mammals (Gumusay et al., 2019). Additionally, acoustics is the primary method of localizing AUVs and underwater GPS beacons. Recent advances in improving localization under environmental uncertainty brings this closer to reality (Graupe et al., 2022). Lastly, new technologies such as Global Navigation Satellite System—Acoustics are revolutionizing the geotechnic community (Yokota et al., 2018).

Distributed acoustic sensing, based on measuring acoustic signals via Rayleigh scattering in fiber optic cables, is revolutionizing investigation of processes operating in the ocean. Cables can either be live (tandem use with commercial data transfer) or dark (dedicated use). Distributed acoustic sensing allows for the investigation of earthquakes, volcanic activity, sediment transport, microseisms, internal and infragravity waves, ocean temperature, mammal vocalizations, wind, and deformation of sea ice and icebergs via noise from events such as calving. In 2014, a 4-day public-domain NSF-funded submarine distributed acoustic sensing experiment was conducted on the two Regional Cabled Array backbone cables (Wilcock et al., 2023). Thirty terabytes of public data were collected, and initial results indicate that acoustic signals were detected from ocean surface waves, ships, Northeast Pacific blue whales, and fin whales (Wilcock et al., 2023), including recording tens of thousands of fin whale calls, and 92 earthquakes (most of which were missing from the seismic databases [Shen and Wu, 2024]). Another 4-day live test, funded by NSF, was successfully conducted in 2024; it is expected that distributed acoustic sensing will be implemented on a continuous basis in the nearshore section of the cable.

SMART Cables

Science Monitoring and Reliable Telecommunications (SMART) cables have the potential to revolutionize ocean data collection; they provide continuous basin-scale ocean measurements in a cost-effective way by instrumenting existing telecommunication infrastructure. A SMART cable includes three simple sensors to measure seafloor motion, water pressure, and temperature (Voosen, 2024). These sensors provide real-time data for earthquake detection, potentially improving early warning of tsunamis. If SMART cables

---

[5] See https://seabed2030.org.

are deployed in sufficient numbers, the measuring of temperature on the seafloor may also be able to indicate how fast deep waters are warming and thus be an important addition to climate studies. The data from SMART cables may also help better understand features in the Earth's mantle (Voosen, 2024). In one of the first deployments of SMART cables worldwide, Portugal is expected to deploy a 3,700-kilometer SMART cable in the Atlantic, connecting mainland Portugal to Madeira and the Azores.

Ocean Biotechnology

Most of the world's biodiversity exists in the ocean. The ocean's plants, animals, and microorganisms display a wide range of biological adaptations and functionality. In the case of microorganisms, this range permits them to live in unusual habitats, including under temperature and pressure extremes. The characteristics of some of these organisms can now be studied using genomic, visualization, and cellular manipulation tools that have been developed for more model systems.

Using genomic techniques, successful applications of biosensing of ocean water, seafloor and subseafloor sediments, and rocks have increased the number of species studied by orders of magnitude, compared with numbers capable of study with ecological methods (i.e., by divers or plankton nets). Whole genomes can be sequenced for a fraction of their costs from a decade ago, and genetic-based probes for specific taxa have been refined with the onset of next-generation sequencing. Entire metabolic pathways—taking raw chemical materials in the ocean and making complex organic products out of them—have been deciphered and recreated. Migration pathways, spawning grounds, rare species, harmful algal blooms, disease agents, fishing impacts, and many other aspects of ocean life are now regularly studied using biotechnology tools.

These advances help the census of ocean life be more accurate across large distances and fall more in line with the kinds of physical and chemical sensors that have long been developed and used in other environments. Yet, the sensing of ocean life is far more intricate, variable, and diverse and requires new development and deployment in the next decade. Some fields are also challenged by interoperability of specific (genetic) technologies to make datasets truly universal and usable across studies. Efforts towards advancing the study of ocean life are needed for the predictive modeling of ocean life to become more on par with the ability to forecast physical and chemical features of ocean water.

Another breakthrough is using marine life itself as a sensor platform. Electronic tags (using satellite, acoustics, etc.) that can be placed on fish or marine mammals are getting smaller and smaller and are capable of higher-resolution detection of key ocean properties. Isotopic profiles of tissues tell a great deal as to where in the food chain a species sits and whether it is feeding on shore or off shore.

Development of Novel Sensors

Sensors are critical for making the sustained observations needed for all three of the report's highlighted science questions. Yet progress on developing novel sensors, especially those that make rate measurements of chemical and biological processes, has lagged behind the development of platforms on which they can be deployed, such as cabled observatories and autonomous platforms. The ocean is a hostile environment, and sensors have to withstand the challenges of corrosion, biofouling, pressure, power, low temperatures, and telemetry. Thus, with a few exceptions, transferring technology developed for nonmarine environments has not been successful. The lack of low-cost, novel sensors is a barrier to observing the 70 percent of the Earth's surface that makes up 95 percent of its biosphere.

While NSF's Ocean Technology and Interdisciplinary Coordination (OTIC) program has been funding sensor innovation in individual investigator laboratories, the development time has been slow, and too few of these novel sensors end up being used by the larger oceanographic communities. There are several contributing causes: (1) sustained funding beyond 3-year grant cycles is lacking, (2) the researcher lacks interest in pursuing the application of their invention beyond their own research needs and their home institution does not incentivize doing so; (3) the researcher is interested in commercialization but does not possess the requisite skills and does not get institutional support; (4) the researcher is successful with a small business innovation research or small business technology transfer (SBIR/STTR) proposal in collaboration with a commercial entity, but the entity is not interested in marketing the invention; (5) the market is not large enough to support the invention at a viable price point; and (6) consolidation in industry reduces competition and innovation. Regardless of the point of failure, the lack of novel sensors has resulted in very slow progress in research beyond the use of sensors for temperature, conductivity, oxygen, pH, light absorption and scatter, and parameters that can be derived from these measurements (e.g., turbulence, fluorescence of phytoplankton pigments and dissolved organic matter, particle size). There have been recent encouraging developments in imaging sensors that can be deployed on autonomous instruments, but most are still too expensive or are constrained by power, weight, and depth capability issues (Boss et al., 2022; Galaz Garcia et al., 2023).

To make the rapid progress necessary to meet the challenges of observing a changing ocean and maintaining the persistent presence needed for improved models, there needs to be a change in the way sensor development is funded, and each of the above points of failure need to be addressed. Opportunities to partner with the Directorate for Technology, Innovation, and Partnerships (TIP) to encourage ocean sensor accelerators that can bring technologies at scale to market within a few years should be further explored. The current approach of SBIR/STTR proposals alone cannot meet this challenge, and innovative methods such as funding instrument pools that can be made available to the community through a competitive proposal process need to be explored. The guaranteed purchase of a large number of instruments might provide the market incentive for commercialization. It should also be noted that not all applications need the same level of accuracy in measurement, and although the two are not mutually exclusive, it is possible that a large number of less accurate sensors may be better suited to meet a science need than a few extremely accurate measurements.

CONCLUSION 4.16: Innovative tools for collecting sustained observations to support the research questions of the next decade are urgently needed. To make the rapid progress necessary to meet the challenges of observing a changing ocean and maintaining the persistent presence needed for improved models, additional modes of funding for, or efforts to change institutional culture regarding, sensor development is needed. Partnership with the NSF TIP Directorate to encourage ocean sensor accelerators is an opportunity that may bring scalable technologies to market within a few years. Regardless of commercial viability, there is a need for fair access to tools for ocean observing: OCE can play a leadership role in ensuring fair access by the oceanographic community.

RECOMMENDATION 4.4: In addition to continuing funding for existing facilities as they support ocean science research priorities for the next decade, the National Science Foundation's Division of Ocean Sciences, in cooperation with agency and other partners (including private philanthropies), should explicitly support:

- **the increased use of autonomous assets for making sustained observations and increasing the efficiency of oceanographic expeditions (expanding the footprint of the vessel while at sea);**
- **new mechanisms for providing researchers with equitable access to instrumentation—for example, shared instrument pools;**
- **the collaborative development of revolutionary and innovative sensor technologies for such efforts as collecting ocean chemistry data (e.g., partial pressure of carbon dioxide), biological data (e.g., environmental and organismal DNA, bio-sensors for species abundance), and measuring seafloor geodesy;**
- **new avenues for bringing novel sensors to market at scale and broadening access to the larger research and management community;**
- **expansion of data curation efforts to support bioinformatics, artificial intelligence, other analyses, and modeling; and the evaluation of new applications of acoustics (e.g., distributed acoustic sensing), Science Monitoring and Reliable Telecommunications cables, and other emerging technologies to identify their potential to contribute to future research.**

CYBERINFRASTRUCTURE

Cyberinfrastructure is an all-encompassing term to describe the necessary hardware, software and procedures to support data acquisition, transmission, curation, storage, protection and distribution, and computer modeling and analysis. As NSF's Directorate for Computer and Information Science and Engineering (CISE) identifies, components of cyberinfrastructure include high performance–computing, software libraries and programming environments, interactive visualization tools, data curation, repositories and management systems, networks, training and protocols for data distribution, outreach and education, and cybersecurity.

Cyberinfrastructure is foundational in facilitating open ocean science through numerical modeling, data analysis, and, most recently, the application of artificial intelligence (AI) and machine learning. Ocean science thrives on data, which are collected at a record and ever-increasing pace. At present, the field of ocean sciences has effective archives and data centers for many of the types of data collected, including at NOAA's National Oceanographic Data Center and National Centers for Environmental Information, the NSF-funded Biological and Chemical Oceanography Data Management Office, the Rolling Deck to Repository for data collected at sea, the Geodetic Facility for the Advancement of Geoscience and the Seismological Facility for the Advancement of Geoscience (GAGE/SAGE), and many others. Many datasets, however, are not curated by national repositories or data centers, and/or the curation process is time consuming and labor intensive, requiring significant expertise and resources. Also, these repositories do not have common data formats and are not well cross-referenced, which is a barrier to AI and machine learning applications. New technologies such as cloud storage and cloud computing offer possibilities that do not require data download, provide high-performance computing access, and offer additional collaboration opportunities. All data need to be governed by FAIR (findable, accessible, interoperable, and reusable) and CARE (collective benefit, authority to control, responsibility, and ethics) principles in order to facilitate transdisciplinary research efforts. Data curation efforts and designs must be driven by the needs of the transdisciplinary users, not only by the requirements of the data archivists, which requires consistent and deep engagement with the diverse array of users.

Ship systems and data stored in the cloud or in online data centers are all vulnerable to future cyberattacks and will require established protections by a skilled workforce (these careers are in demand in all sectors). For example, having information technology and Efforts to Outcomes specialists on board all UNOLS vessels is essential to keeping the ARF safe. While this increases vessel operating day rates, the importance of these positions on board cannot be overemphasized.

Because the details of these complex considerations exceeded the committee's expertise, it did not fully evaluate cyberinfrastructure needs. However, several national and international efforts have developed strategies for FAIR and CARE data management.

One of the themes for the U.S. Contribution to the UN Decade of Ocean Science for Sustainable Development is the "Ocean of Data" (NASEM, 2022b), which lays out plans to conduct convening activities to determine best practices for data curation, accessibility and protection. Additionally, the Ocean Research Advisory Panel's (2024) *Toward a National Ocean Data Strategy*, proposes a strategy for improving data management, growing partnerships, and advancing access and usability. The UN Ocean Decade Data and Information working group, made of many U.S. members, has released a strategy for establishing a trusted, comprehensive, and interconnected ocean data and information system. In recognition of the complex issue of cyberinfrastructure across the scientific community, NSF has recently established the Office of Advanced Cyberinfrastructure (OAC) within CISE to understand these issues across the science community. These activities may be helpful to OCE and other agencies when considering best practices and ways to advance data practices.

Improvements in all areas of cyberinfrastructure and cyber-related applications, including data curation, uncertainty quantification, AI, numerical modeling, FAIR and CARE principles, cybersecurity at sea and on land, and other aspects are essential components of successfully addressing all three of the science questions highlighted in this report. The committee acknowledges that these efforts are not inexpensive and require careful consideration.

CONCLUSION 4.17: The collection and effective integration of disparate ocean datasets, especially new datasets from emerging technologies, such as biotechnology to support AI and machine learning, are needed to create a unified, accessible, and interoperable ocean information system. Much can be gleaned from existing and recent national and international efforts to develop robust ocean data management strategies; a workshop or other convening activity would be an appropriate next step for OCE.

RECOMMENDATION 4.5: The National Science Foundation's Division of Ocean Sciences should fund a convening activity, such as a series of workshops, that seeks to gather expert advice and input, review established strategies, and develop peer-reviewed guidelines and practices for ocean science data curation, computing, and security, both on research vessels and on shore, integrating findable, accessible, interoperable, and reusable (FAIR) and collective benefit, authority to control, responsibility, ethics (CARE) principles and the required supporting cyberinfrastructure. Data and computational experts from adjacent fields (e.g., applied mathematics) should be encouraged to participate.

5

A New Decade of Ocean Science Research

The new decade will continue to see transformations in the world's ocean, and thus ocean science research must also transform, using new technologies, increasing the ability to forecast future changes, and improving benefits to society. Both basic and use-inspired research are needed to meet the anticipated challenges. New knowledge from focused disciplinary research needs to be integrated through transdisciplinary approaches leading to a holistic understanding of the ocean and improved forecasts. Data and information gained will need to incorporate a broad base of knowledge, including from traditional sources, Indigenous and local knowledge, collaborative development and production of research and knowledge, and citizen science. Workforce development and education will be key to preparing and enhancing the multifaceted talent pool that will be needed to support and advance oceanographic research. Expanded and fairly distributed investments in research, infrastructure, technology, and workforce will fuel increased understanding and improve the capability to forecast changes.

The coming decade will see a more rapid evolution toward an integrated, collaborative, and technologically advanced approach to understanding the world's ocean, as well as progressing the capability to forecast the future ocean at a variety of scales relevant to human life. Bridging the gap between disciplines while simultaneously leveraging cutting-edge, discipline-based research methodologies and technologies will enhance understanding of the interconnected nature of physics, chemistry, biology, and geology, and address the multiple stressors facing the ocean. For example, the nature-based solution of using oyster reefs to protect shorelines from storm surges requires integration of fundamental understanding of fine-scale hydrodynamics, sediment transport, ocean chemistry, phytoplankton primary production, and oyster physiology, as well as various aspects of societal needs. These approaches recognize the complexity and interdependency of the ocean system. The outcomes of this new research portfolio will deepen understanding and support innovative and cost-effective solutions for mitigating and adapting to the impact of changes to the Earth system on ocean processes and coastal communities.

Rapid and timely curation, processing, and analysis of the vast amounts of ocean data that are being collected will be required to support accurate and reliable forecasts. The use of artificial intelligence (AI), machine learning, and cyberinfrastructure is rapidly advancing, promising to enhance real-time analysis of data and ocean model output. Acquiring adequate resources critical for taking full advantage of new data sources and processing technologies and to sustain measurements into the future will require the National Science Foundation's (NSF's) Division of Ocean Sciences (OCE) to forge new relationships and foster collaboration with international partners, philanthropic foundations, and private-sector companies.

The committee hopes this report will galvanize the U.S. and international ocean science community around the common global goal and challenge of forecasting critical ocean processes on scales that matter most to humankind, and that, by the end of the next decade, the science, tools, and technology needed to tackle the urgent research priorities in ocean and earth sciences will be available. Led by the OCE, and together with other NSF directorates and divisions, federal agencies, and nongovernmental organizations, the headlines in 2035 could be quite different from today's. Headlines like "NSF Announces Bold new Partnership to Sample the Ocean's Depths," "Earthquake Early Warning System Saves Lives and Livelihoods," and "Predictions of the Evolution of the AMOC System now Available Through 2045—How to Prepare for What is to Come" could be realized.

The committee sought balance between goals possible under different funding realities as well as opportunites to fill gaps by leveraging resources and building strategic partnerships. This chapter is focused on providing a framework for OCE to create a roadmap to address the urgent science needed and prioritize the enabling elements, such as research vessels and laboratories, ocean observations, an expanded and broadened workforce, coordination of data, computing centers, and new partnerships. Strong U.S. leadership, as well as international collaboration, is necessary to address urgent ocean science needs via these enabling elements. These points are further emphasized in the scientific motivation behind the establishment of the 5th International Polar Year, which was developed simultaneously with this report, prioritizing overlapping research priorities, infrastructure needs, and transdisciplinary approaches to research (Box 5.1).

BOX 5.1
The Polar Oceans: An "All Hands on Deck" Example

It is striking that the scientific motivation for the establishment of a 5th International Polar Year (IPY5, n.d.), 2032–2033 is so similar to the global ocean science priorities, needs, and approaches presented in this 2025 Decadal Survey report. While independently written, both recognize the potential impacts to economies, ecosystems, and human well-being of increasing temperatures, sea level rise, ocean acidification, changes in sea ice, and the possibility of extreme events. Both recognize that some rates of change in the Earth system and its ocean system are occurring rapidly with the potential for complex nonlinear responses and feedbacks. And both recognize that many of these changes are linked to processes occurring in the Arctic and Antarctic polar regions. For example, the polar and subpolar regions are where all deep water forms (sinks) at the ocean surface, and thus these regions are a first-order control on global ocean circulation, the related redistribution of heat and salt, and the supply of dissolved oxygen to the marine realm. Ice sheet–ocean interactions (Figure 5.1, lower) at both poles affect the formation of deep water. Oceanic field work and remote sensing, as well as the collection of marine records of paleo ice sheet and paleo ocean behaviors are important for supplying needed data to inform and test predictive models on future sea level rise, rates of change, and ecosystem responses to changing boundary conditions. Additionally, changing sea ice conditions in the Arctic have a direct impact on marine food webs, potential access to seabed resources, global commerce (Figure 5.1, upper), and U.S. national security. These changes will likely lead to greater competition for power in the region. Improving scientific understanding of the Arctic environment is one strategy to prepare for changes to come.

FIGURE 5.1 Aerial view of a container ship in icy waters of the Arctic Ocean (upper) and scientific research vessel *Polarstern* conducting research in the Southern Ocean adjacent to the floating ice shelf of the West Antarctic Ice Sheet (lower).
SOURCE: Upper: Thomas Ronge, International Ocean Discovery Program; lower: A. Medvedkov, Arctic Council, https://arctic-council.org/projects/arctic-ship-traffic-data-astd.

Investments in polar ocean science will need funding both for infrastructure and basic/applied research. This is especially true given the reduction in the number and type of polar research vessels (e.g., vessels that can operate in polar open waters, ice-reinforced vessels, and icebreakers), which if not addressed, makes it harder for the United States to make meaningful contributions to polar ocean science. Such infrastructure needs were highlighted in the recent report *Future Directions for Southern Ocean and Antarctic Nearshore and Coastal Research* (NASEM, 2024d) and the letter report, *Acquisition and Operation of Polar Icebreakers: Fulfilling the Nation's Needs* (NASEM, 2017b). Additionally, increased research on nonlinear dynamics and support for sustained, long-term observing to distinguish directional changes in polar oceans are needed.

Both the IPY5 planning and recommendations in this Decadal Survey report identify the need for transdisciplinary coordinated efforts. Especially in the Arctic, such ocean science efforts can benefit from the long history of Indigenous knowledge. Indigenous well-being is also a foundational principle to guide Arctic Ocean research. Transdisciplinary coordination will require collaboration and partnerships between OCE and NSF's Office of Polar Programs, with multisector U.S. agencies, such as the National Oceanic and Atmospheric Administration and the Department of Defense, and with other countries and international polar organizations (e.g., Arctic Council). U.S. scientific leadership in polar ocean science is perhaps never more important than in this time of global transitions. U.S. leadership in multilateral science and technology forums is one way to strengthen the U.S. position to compete and collaborate internationally (Subcommittee on International Science & Technology Coordination [ISTC], 2024).

INTEGRATED FORECASTS IN OCEAN SCIENCE

As explained in Chapter 2, the committee presents a challenge for the coming decade: by 2035, establish a new paradigm for forecasting ocean processes at scales relevant to human well-being. This challenge is based on the following:

- The ocean is changing in unanticipated ways and at an unprecedented pace.
- Understanding and anticipating the change in the ocean has never been more urgent.
- To truly address these changes to the ocean, basic research in key areas is needed to advance predictive technologies.

The interconnected nature of the urgent ocean science research portfolio emphasizes the need to study the ocean as part of the entire Earth system. Research to improve understanding of the function and role of the ocean in the Earth system need to continue, in addition to employing transdisciplinary research approaches to fully understand important aspects of the ocean and to enable forecasts of future change. For example, a study of ocean heat and carbon transport will support a focus on the interconnections between these three urgent research themes (see Box 2.2 in Chapter 2) by considering the fundamental role of microorganisms in marine food webs and in carbon transformation. Direct measurements and models of such interconnections, which are largely missing from ecosystem models today, will support a better understanding of feedbacks[1] within these three core themes and their impacts within the wider Earth system. Studying marine heat waves as holistic phenomena will provide greater understanding of their impact on the atmosphere, on ocean mixing, on the exchange of carbon and nutrients between surface and deeper waters, on primary productivity, on organismal physiology, and on ecosystem connections. Even further, studying heat and carbon transport can also help forecast the health of corals, which have direct impact on coastal effects from storm surge (Box 5.2). The committee regards the list of examples that can be derived from these interconnections infinitely generative; clearly understanding the interconnected ocean system will be key to tackling the three urgent ocean science research questions. This is further emphasized by scientific motivations, priorities, needs, and approaches being pursed for polar ocean research (Box 5.1).

---

[1] A *feedback* is defined as a process working within a system that can either increase or decrease the effects of change, and in this case, climate change.

BOX 5.2
Coral Health: Interconnected Nature of Ocean Sciences

FIGURE 5.2 Shallow-water coral reef supporting an array of marine life at Palmyra Atoll National Wildlife Refuge. SOURCE: U.S. Fish and Wildlife Service.

Humans living in coastal communities have deep generational knowledge about the ocean and health. Historically, as scientists have gained more knowledge about the ocean, integration of generational knowledge and multiple knowledge systems have been largely ignored. Coral reefs are an excellent example of an ecosystem that requires transdisciplinary research, with inclusion of historical and biocultural knowledge, to advance understanding. Corals build reefs of carbonate are among the largest biological structures on Earth; these reefs not only have the highest biodiversity of any ecosystem on Earth but are a critical food source, serve as a vault for sequestering carbon dioxide, and protect coastal communities from the effects of extreme storms. True advancement through transdisciplinary research would include investing in improving the ability to forecast the health of these important systems, through integration of disciplines and sectors such as social science, engineering, biophysical research, and computer science. In addition, revival of subsistence fishing practices within many coastal communities and understanding of human dimensions and cultural ecosystem services through transdisciplinary approaches will ensure ecosystem resilience through effective management of resources.

Storm damage from rising sea levels is mitigated by reefs comprised of coastal species such as corals, oysters, mangroves, and marshes, which act to slow rough wave action and mitigate onshore disasters. Changes in temperature also impact ecosystems. Ocean temperatures that exceed the tolerance threshold of organisms can cause certain species to migrate to cooler waters, thus creating shifts in fishing habitats, changes in shipping lanes to avoid whale strikes, changes in the species fished, and sometimes reconsideration of marine protected areas.

To truly grasp the ocean's role in the Earth system and forecast the future health of the planet requires ocean sciences research that integrates all three priority themes: ocean and climate, ecosystem resilience, and extreme events.

Key to ensuring integration across these themes is the establishment of a cross-cutting, comprehensive oceanography program that has as its sole purpose linking the inherent understanding, and hence predictions, from each of the three themes. Precedents exist for allocating new resources to support such an integrative program, but a large-scale, coordinated effort to which NSF makes a long-standing commitment is now critically needed to address the societal needs arising from the challenges faced. Such an effort would require new investment involving significant coordination and partnerships with other directorates, organizations, and agencies, as well as other disciplines. Any such program would not only integrate across the themes but also provide an opportunity to support a sense of community across all oceanographic disciplines that is typically lacking.

CONCLUSION 5.1: To truly accelerate progress in ocean science, new knowledge will need to be integrated into research projects focused across the three priority themes: ocean and climate, ecosystem resilience, and extreme events. A cross-cutting, comprehensive, and sustainable program could be established with the sole purpose of connecting the three themes. Enabling this connection requires significant resources, coordination, and established partnerships with multiple NSF directorates, organizations, and agencies, as well as other disciplines. NSF-wide mechanisms such as existing and new Science and Technology Centers and Long-Term Ecological Research programs could be leveraged to achieve the NSF-wide support for such a program.

REGAINING U.S. LEADERSHIP IN OCEAN SCIENCE

OCE-supported research at academic and other institutions has made the United States the world leader in ocean science research, leading to many scientific advances and understanding ocean processes in all subfields of the ocean sciences. However, the rest of the world is now also investing heavily in research in the ocean sciences; large-scale science projects, such as the scientific ocean drilling program and the Argo program, depend upon the talent and resources of multiple countries working together. In the next decade, the United States has the opportunity to strengthen existing partnerships with other countries, build new partnerships, and develop procedures for evaluating the success of these collaborations.

The next decade of ocean science research is critical, and without significant new investment, U.S. leadership will continue to erode, resulting in a loss of expertise and talent to other countries. A loss of worldwide leadership in science, technology, engineering, and mathematics (STEM) impacts the U.S. economy and national security. The committee believes that consideration of conclusions and full implementation of recommendations included in this report will be important steps toward regaining NSF-supported U.S. leadership in ocean science research. Demonstrating strong U.S. leadership in ocean science research will also improve management and decision-making of our coastal and ocean resources, enhance national security and economic prosperity, and propel the blue economy in burgeoning sectors of marine carbon dioxide removal, harmful algal bloom mitigation, development of aquaculture, and other applications.

CONCLUSION 5.2: The next decade of ocean science research is critical, and without significant new investment, U.S. leadership will continue to erode. Demonstrating strong U.S. leadership in ocean science research will improve management and decision-making of our coastal and ocean resources, enhance national security and economic prosperity, and propel the blue economy.

IMPLEMENTATION AMID UNCERTAINTY

Without knowledge of future budgets and costs to implement the committee's recommendations, it is not possible to lay out a full implementation plan for the next decade. Some recommendations can likely be implemented within a level budget or with modest increases; others will require substantial new resources. For example, significant and strategic investments will be required for new initiatives for sensors, AI and machine learning, real-time hazard monitoring, numerical models, data curation, and replacement of current facilities.

Overarching Framework

NSF should continue to fund basic research proposals, with a balanced portfolio across the subdisciplines of all ocean sciences (Recommendation 2.1), and with special consideration of the three science questions the committee poses:

- Ocean and Climate—How will the ocean's ability to absorb heat and carbon change?
- Ecosystem Resilience—How will marine ecosystems respond to changes in the Earth system?
- Extreme Events—How can the ability to forecast extreme events driven by ocean and seafloor processes be improved?

This research should emphasize advancing forecasting with deep understanding of ocean processes (Recommendation 2.2), while also creating opportunities to foster fair and indiscriminate transdisciplinary research practices, as appropriate (Recommendation 3.1). Implementation of Recommendation 3.1 also requires complementary action on the part of OCE's academic partners.

Immediate Actions for Opportunities and Impacts

A subset of the recommendations in this report will require time to mature, so work to advance the recommendations will benefit from immediate action in order to fully support the innovative, societally relevant, and use-inspired urgent ocean and Earth science research of the next decade. Efforts that can be, and must be started immediately are as follows:

- Investments to expand and broaden participation in the ocean sciences with transdisciplinary teams—the science required for national success cannot be accomplished without a larger workforce that includes broad representation (Recommendation 3.2).
- Obtain ocean observations that will advance the urgent ocean science research questions, and to include strategic partnerships (Recommendation 3.3), such as with the Technology, Innovation and Partnership Directorate, to develop and produce ocean observing technologies critical to achieving the science of the next decade (Recommendation 4.4).
- Support and enhance data curation and thoughtful consideration of ocean data stewardship, in part so it is most useful for sophisticated analyses and to support improved numerical models (Recommendation 4.5).

Addressing Infrastructure Needs

The next subset of the recommendations is related to the infrastructure needed to support research related to the three urgent ocean science questions. Planning needs to begin now to reap benefits within the next decade.

- Continued and enhanced operation of the Academic Research Fleet, taking into account the contributions of autonomous platforms that can expand the footprint of vessels at sea and the replacement schedule for the aging fleet (Recommendation 4.1).
- Continued operation of components of the Ocean Observatories Initiative that are required to implement the committee's science questions but with a timely review of the science outcomes and careful consideration of its fundamental scientific framework (Recommendation 4.2).
- Enabling future scientific ocean drilling in support of the urgent research questions, including planning to replace what was lost with decommissioning the *JOIDES Resolution* (Recommendation 4.3).

- Continued development, evaluation, and implementation of new technologies and approaches, as well as other innovations that may arise in the coming decade (Recommendation 4.4).
- Work with other directorates, agencies, the private sector, and philanthropic organizations, perhaps by hosting convening sessions, to learn how to take full advantage of advances in cyberinfrastructure to support accelerated science and discovery (Recommendation 4.5).

In summary, the challenge in the field of ocean sciences for the next decade has been constructed on a foundation of an expanded and broadened workforce, as well as utilization of existing and new infrastructure (Figure 5.3). Answering the urgent ocean science research questions laid out in this report will necessitate individual principal investigator–driven research, as well as interdisciplinary and transdisciplinary research teams working together. Achieving the desired forecasting capabilities will result from an iterative process combining advanced and sustained ocean observations, modeling, the use of artificial intelligence and machine learning, and in-depth scientific understanding. Achieving this will also require additional resources and new partnerships with other NSF directorates, other agencies, philanthropies, industry, and international organizations.

FIGURE 5.3 From basic research to forecasts of ocean processes at human scales.

MEASURING OCE'S SUCCESS

Moving forward, periodic reviews of large-scale programs supported by OCE (e.g., the Academic Research Fleet, Ocean Observatories Initiative, and Ocean Drilling Program), will determine the value of these programs to the scientific community, to the funders, and to the public. These reviews need to go beyond evaluating how well the infrastructure is meeting cooperative agreement deliverables and include an assessment of the scientific and societal outcomes enabled by infrastructure investments. Reviews of this nature also need to allow for feedback and mechanisms for course correction.

Similarly, at the mid-decade, a self-review of OCE's progress towards implementing the 2025 Decadal Survey recommendations is encouraged to serve as a tool for communicating progress to the public and other partners, providing insight into the specific downstream effects of the OCE budget environment and valuable feedback into the development of any future decadal survey of ocean sciences. Such a review could be led by the Advisory Committee for Geosciences and with input from the broader ocean and Earth science community.

RECOMMENDATION 5.1: Any large-scale, long-term program (e.g., Academic Research Fleet, Ocean Observatories Initiative, Ocean Drilling Program) supported by the National Science Foundation (NSF) should be periodically evaluated on how the program is achieving its science goals and meeting the science community needs and to allow for feedback and mechanisms to correct course. Evaluations should include meaningful metrics that measure the scientific and societal benefits, including community engagement and workforce training, not just whether the given program is meeting its operational objectives. Outcomes of the evaluations should be shared with NSF and the public. Similarly, after 5 years (i.e., the mid-point of the decade), the committee recommends that the Division of Ocean Sciences (OCE) review and assess its progress towards implementing the 2025 Decadal Survey recommendations, with consideration to the broader OCE budget environment.

A NEW DECADE, A NEW APPROACH

With the goal of realizing the challenge of the next decade, the committee approached the statement of task (see Box 1.2 in Chapter 1) with a lens of doing science differently. Equal to defining the research portfolio and infrastructure needed for the next decade of ocean research and beyond is the importance of broadening participation and creating opportunities to cross societal boundaries, disciplines, and sectors. In Box 5.3, the committee presents its call to action to the broader ocean science community to tackle the complex and urgent challenges of the coming decade described in this report.

BOX 5.3
A Call to Action for Ocean Sciences

Progress in ocean sciences research over the next decade will require a directed effort to establish predictive capabilities and enable accurate forecasts of the interconnected ocean processes at scales relevant to human life. This report calls for an "all hands on deck" approach to ocean science: embracing new and evolving tools and methodologies, establishing transdisciplinary ocean science research teams, and integrating Indigenous and local knowledge systems into the research process. By addressing the human dimensions within the field of ocean science, along with advancing observations and predictions, new discoveries and knowledge will be made that enhance adaptation, resiliency, and prosperity of humankind and the ecosystems foundational for human life and well-being.

The committee urges the greater ocean sciences community to thoughtfully consider initiatives and pathways that promote equal opportunity in workforce training, as well as provide access and agency in data management and usage. The committee champions the importance of fundamental science approached through a transdisciplinary lens and implementation of mechanisms that will allow the research results to be rapidly shared and utilized.

Furthermore, the committee strongly encourages members of the existing ocean science community (along with new partners and interest holders) to bring their talents and skills to the specific questions outlined herein—to innovate, collaborate, and push the boundaries of knowledge. In order to maximize the impact of future scientific endeavors, the committee strongly encourages explicit assessment of progress while cultivating fertile ground for unforeseen, groundbreaking discoveries. By integrating knowledge and expertise across disciplines and communities, it can be ensured that ocean science research remains relevant and continues to improve the well-being of our planet and all its inhabitants.

CONCLUSION

OCE-supported research at academic and other institutions has helped make the United States the world leader in ocean science research, leading to many scientific advances and understanding of ocean processes in all subfields within the field of ocean sciences. This new decade is starting out at a time when U.S. investment in STEM in general, and the geosciences in particular, is not keeping pace with growing societal needs, including better understanding of the ecosystems supporting fisheries, changes in the Arctic related to national security, and the capacity of the ocean to absorb heat from a warming atmosphere. At the same time, scientific leadership and competition from other countries, including U.S. competitors, are increasing. The U.S. investments in ocean science in the last decade do not take into account the changing landscape of ocean science research today or the needs of the future. Thus, the past 10 years of funding is not representative of what is needed in the next 10 years. The United States is at a critical juncture in which major investments are needed for new and replacement infrastructure, and to support basic and applied research, without which the United States may be left behind in global ocean science research transformation.

The 2025 Decadal Survey was developed to ensure that OCE will continue to provide the foundation for contributions to ocean science research, not only within NSF but also across the federal agencies that use research in ocean science to accomplish their mission. Now is the time for the United States to invest and take leadership in answering the urgent ocean science priorities outlined in this report for a healthy and well-understood ocean, allowing the anticipation and assessment of changes, which is critical for national security, economic prosperity, and environmental stewardship, and for the well-being of humans and the ecosystems on which people depend, in the next decade and beyond.

References

Adger, W. N., J. Paavola, and S. Huq. 2006. "Toward justice in adaptation to climate change." In *Fairness in Adaptation to Climate Change*, edited by W. Neil Adger, Jouni Paavola, Saleemul Huq, and M. J. Mace. Cambridge: The MIT Press.

Ait Brahim, Y., M. C. Peros, A. E. Viau, M. Liedtke, J. M. Pajón, J. Valdes, X. Li, R. L. Edwards, E. G. Reinhardt, and F. Oliva. 2022. "Hydroclimate variability in the Caribbean during North Atlantic Heinrich cooling events (H8 and H9)." *Scientific Reports* 12(1):20719. https://doi.org/10.1038/s41598-022-24610-x.

Ali, H. N., S. L. Sheffield, J. E. Bauer, R. P. Caballero-Gill, N. M. Gasparini, J. Libarkin, K. K. Gonzales, J. Willenbring, E. Amir-Lin, and J. Cisneros. 2021. "An actionable anti-racism plan for geoscience organizations." *Nature Communications* 12(1):1-6.

Allison, S. D., and J. B. H. Martiny. 2008. "Resistance, resilience, and redundancy in microbial communities." *Proceedings of the National Academy of Sciences* 105(supplement_1):11512-11519. https://doi.org/10.1073/pnas.0801925105.

AlShebli, B. K., T. Rahwan, and W. L. Woon. 2018. "The preeminence of ethnic diversity in scientific collaboration." *Nature Communications* 9(1):5163. https://doi.org/10.1038/s41467-018-07634-8.

Aminpour, P., S. A. Gray, A. Singer, S. B. Scyphers, A. J. Jetter, R. Jordan, R. Murphy, and J. H. Grabowski. 2021. "The diversity bonus in pooling local knowledge about complex problems." *Proceedings of the National Academy of Sciences* 118(5):e2016887118. https://doi.org/doi:10.1073/pnas.2016887118.

Amon, D. J., S. Gollner, T. Morato, C. R. Smith, C. Chen, S. Christiansen, B. Currie, J. C. Drazen, T. Fukushima, M. Gianni, K. M. Gjerde, A. J. Gooday, G. G. Grillo, M. Haeckel, T. Joyini, S.-J. Ju, L. A. Levin, A. Metaxas, K. Mianowicz, T. N. Molodtsova, I. Narberhaus, B. N. Orcutt, A. Swaddling, J. Tuhumwire, P. U. Palacio, M. Walker, P. Weaver, X.-W. Xu, C. Y. Mulalap, P. E. T. Edwards, and C. Pickens. 2022. "Assessment of scientific gaps related to the effective environmental management of deep-seabed mining." *Marine Policy* 138:105006. https://doi.org/10.1016/j.marpol.2022.105006.

Amon, D. J., B. R. C. Kennedy, K. Cantwell, K. Suhre, D. Glickson, T. M. Shank, and R. D. Rotjan. 2020. "Deep-sea debris in the Central and Western Pacific Ocean." *Frontiers in Marine Science* 7. https://doi.org/10.3389/fmars.2020.00369.

Anderson, D. M., E. Fensin, C. J. Gobler, A. E. Hoeglund, K. A. Hubbard, D. M. Kulis, J. H. Landsberg, K. A. Lefebvre, P. Provoost, M. L. Richlen, J. L. Smith, A. R. Solow, and V. L. Trainer. 2021. "Marine harmful algal blooms (HABs) in the United States: History, current status and future trends." *Harmful Algae* 102:101975. https://doi.org/10.1016/j.hal.2021.101975.

Anderson, L. B., B. Hönisch, H. K. Coxall, and L. Bolge. 2024. "Atmospheric CO estimates for the Late Oligocene and Early Miocene using multi-species cross calibrations of boron isotopes." *Paleoceanography and Paleoclimatology* 39(1):e2022PA004569. https://doi.org/ARTN e2022PA0045 6910.1029/2022PA004569.

Archibald, M. M., M. T. Lawless, M. A. P. de Plaza, and A. L. Kitson. 2023. "How transdisciplinary research teams learn to do knowledge translation (KT), and how KT in turn impacts transdisciplinary research: A realist evaluation and longitudinal case study." *Health Research Policy and Systems* 21(1):20. https://doi.org/10.1186/s12961-023-00967-x.

Armbrecht, L., M. E. Weber, M. E. Raymo, V. L. Peck, T. Williams, J. Warnock, Y. Kato, I. Hernández-Almeida, F. Hoem, B. Reilly, S. Hemming, I. Bailey, Y. M. Martos, M. Gutjahr, V. Percuoco, C. Allen, S. Brachfeld, F. G. Cardillo, Z. Du, G. Fauth, C. Fogwill, M. Garcia, A. Glüder, M. Guitard, J.-H. Hwang, M. Iizuka, B. Kenlee, S. O'Connell, L. F. Pérez, T. A. Ronge, O. Seki, L. Tauxe, S. Tripathi, and X. Zheng. 2022. "Ancient marine sediment DNA reveals diatom transition in Antarctica." *Nature Communications* 13(1):5787. https://doi.org/10.1038/s41467-022-33494-4.

Bammer, G., C. A. Browne, C. Ballard, N. Lloyd, A. Kevan, N. Neales, T. Nurmikko-Fuller, S. Perera, I. Singhal, and L. van Kerkhoff. 2023. "Setting parameters for developing undergraduate expertise in transdisciplinary problem solving at a university-wide scale: a case study." *Humanities and Social Sciences Communications* 10(1):208. https://doi.org/10.1057/s41599-023-01709-8.

Bar-On, Y. M., R. Phillips, and R. Milo. 2018. "The biomass distribution on Earth." *Proceedings of the National Academy of Sciences* 115(25):6506-6511. https://doi.org/10.1073/pnas.1711842115.

Bay, R. A., N. H. Rose, C. A. Logan, and S. R. Palumbi. 2017. "Genomic models predict successful coral adaptation if future ocean warming rates are reduced." *Science Advances* 3(11):e1701413. https://doi.org/10.1126/sciadv.1701413.

Bayliss-Brown, G., L. Cavaleri Gerhardinger, and C. Starger. 2020. "Networked knowledge to action in support of ocean sustainability." *Coastal Management* 48(4):235-237. https://doi.org/10.1080/089207 53.2020.1778426.

BEA (Bureau of Economic Analysis). 2020. *Defining and Measuring the U.S. Ocean Economy* https://www.bea.gov/system/files/2021-06/defining-and-measuring-the-united-states-ocean-economy.pdf (accessed March 20, 2025).

BEA. 2024. *Marine Economy Satellite Account, 2022.* https://www.bea.gov/news/2024/marine-economy-satellite-account-2022 (accessed March 20, 2025).

Beadling, R. L., J. P. Krasting, S. M. Griffies, W. J. Hurlin, B. Bronselaer, J. L. Russell, G. A. MacGilchrist, J.-E. Tesdal, and M. Winton. 2022. "Importance of the Antarctic Slope Current in the Southern Ocean Response to ice sheet melt and wind stress change." *Journal of Geophysical Research: Oceans* 127(5):e2021JC017608. https://doi.org/10.1029/2021JC017608.

Behl, M., S. Cooper, C. Garza, S. E. Kolesar, S. Legg, J. C. Lewis, L. White, and B. Jones. 2021. "Changing the culture of coastal, ocean, and marine sciences strategies for individual and collective actions." *Oceanography* 34(3):53-60. https://www.jstor.org/stable/27051390.

Bendor, J., and S. E. Page. 2019. "Optimal team composition for tool-based problem solving." *Journal of Economics & Management Strategy* 28(4):734-764. https://doi.org/10.1111/jems.12295.

Berkes, F. 2003. "Can cross-scale linkages increase the resilience of social-ecological systems?" Religious Communities and Sustainable Development International Conference, Politics of the Commons, Chiang Mai, Thailand.

Bishop, M. J., M. Mayer-Pinto, L. Airoldi, L. B. Firth, R. L. Morris, L. H. L. Loke, S. J. Hawkins, L. A. Naylor, R. A. Coleman, S. Y. Chee, and K. A. Dafforn. 2017. "Effects of ocean sprawl on ecological connectivity: Impacts and solutions." *Journal of Experimental Marine Biology and Ecology* 492:7-30. https://doi.org/10.1016/j.jembe.2017.01.021.

Blackwell, E., M. Shirzaei, C. Ojha, and S. Werth. 2020. "Tracking California's sinking coast from space: Implications for relative sea-level rise." *Science Advances* 6(31):eaba4551. https://doi.org/10.1126/sci adv.aba4551.

Borenstein, S. 2024. "The hottest summer on record could lead to our warmest year ever measured. Again." PBS, Sept 6. https://www.pbs.org/newshour/science/the-hottest-summer-on-record-could-lead-to-our-warmest-year-ever-measured-again (accessed March 20, 2025).

Boss, E., A. M. Waite, J. Karstensen, T. Trull, F. Muller-Karger, H. M. Sosik, J. Uitz, S. G. Acinas, K. Fennel, I. Berman-Frank, S. Thomalla, H. Yamazaki, S. Batten, G. Gregori, A. J. Richardson, and R. Wanninkhof. 2022. "Recommendations for Plankton Measurements on OceanSITES Moorings with Relevance to Other Observing Sites." *Frontiers in Marine Science* 9. https://doi.org/10.3389/fmars.2022. 929436.

Boyd, P. W., H. Claustre, M. Levy, D. A. Siegel, and T. Weber. 2019. "Multi-faceted particle pumps drive carbon sequestration in the ocean." *Nature* 568(7752):327-335. https://doi.org/10.1038/s41586-019-1098-2.

Bradley, A. T., and I. J. Hewitt. 2024. "Tipping point in ice-sheet grounding-zone melting due to ocean water intrusion." *Nature Geoscience* 17(7):631-637. https://doi.org/10.1038/s41561-024-01465-7.

Breslow, S. J., B. Sojka, R. Barnea, X. Basurto, C. Carothers, S. Charnley, S. Coulthard, N. Dolsak, J. Donatuto, C. García-Quijano, C. C. Hicks, A. Levine, M. B. Mascia, K. Norman, M. Poe, T. Satterfield, K. St Martin, and P. S. Levin. 2016. "Conceptualizing and operationalizing human wellbeing for ecosystem assessment and management." *Environmental Science & Policy* 66:250-259. https://doi.org/10.1016/j.envsci.2016.06.023.

Bronselaer, B., M. Winton, S. M. Griffies, W. J. Hurlin, K. B. Rodgers, O. V. Sergienko, R. J. Stouffer, and J. L. Russell. 2018. "Change in future climate due to Antarctic meltwater." *Nature* 564(7734):53-58. https://doi.org/10.1038/s41586-018-0712-z.

Brose, U., P. Archambault, A. D. Barnes, L. F. Bersier, T. Boy, J. Canning-Clode, E. Conti, M. Dias, C. Digel, A. Dissanayake, A. A. V. Flores, K. Fussmann, B. Gauzens, C. Gray, J. Haussler, M. R. Hirt, U. Jacob, M. Jochum, S. Kefi, O. McLaughlin, M. M. MacPherson, E. Latz, K. Layer-Dobra, P. Legagneux, Y. Li, C. Madeira, N. D. Martinez, V. Mendonca, C. Mulder, S. A. Navarrete, E. J. O'Gorman, D. Ott, J. Paula, D. Perkins, D. Piechnik, I. Pokrovsky, D. Raffaelli, B. C. Rall, B. Rosenbaum, R. Ryser, A. Silva, E. H. Sohlstrom, N. Sokolova, M. S. A. Thompson, R. M. Thompson, F. Vermandele, C. Vinagre, S. Wang, J. M. Wefer, R. J. Williams, E. Wieters, G. Woodward, and A. C. Iles. 2019. "Predator traits determine food-web architecture across ecosystems." *Nature Ecology & Evolution* 3(6):919-927. https://doi.org/10.1038/s41559-019-0899-x.

Bugnot, A. B., M. Mayer-Pinto, L. Airoldi, E. C. Heery, E. L. Johnston, L. P. Critchley, E. M. A. Strain, R. L. Morris, L. H. L. Loke, M. J. Bishop, E. V. Sheehan, R. A. Coleman, and K. A. Dafforn. 2021. "Current and projected global extent of marine built structures." *Nature Sustainability* 4(1):33-41. https://doi.org/10.1038/s41893-020-00595-1.

Bush, V. 1945. *Science: The Endless Frontier.* A Report to the President by Vannevar Bush, Director of the Office of Scientific Research and Development, July 1945. Washington, DC: U.S. Government Printing Office.

Bushinsky, S. M., P. Landschutzer, C. Rodenbeck, A. R. Gray, D. Baker, M. R. Mazloff, L. Resplandy, K. S. Johnson, and J. L. Sarmiento. 2019. "Reassessing Southern Ocean air-sea CO_2 flux estimates with the addition of biogeochemical float observations." *Global Biogeochemical Cycles* 33(11):1370-1388. https://doi.org/10.1029/2019GB006176.

Buzzanga, B., D. P. S. Bekaert, B. D. Hamlington, R. E. Kopp, M. Govorcin, and K. G. Miller. 2023. "Localized uplift, widespread subsidence, and implications for sea level rise in the New York City metropolitan area." *Science Advances* 9(39):eadi8259. https://doi.org/doi:10.1126/sciadv.adi8259.

Caínzos, V., A. Velo, F. F. Pérez, and A. Hernández-Guerra. 2022. "Anthropogenic carbon transport variability in the Atlantic Ocean over three decades." *Global Biogeochemical Cycles* 36(11):e2022GB007475. https://doi.org/10.1029/2022GB007475.

Camoin, G., and N. Eguchi. 2024. "The International Ocean Drilling Programme (IODP3)." *Scientific Drilling* 33(1):89-92. https://doi.org/10.5194/sd-33-89-2024.

Cangialosi, J. P., E. Blake, M. DeMaria, A. Penny, A. Latto, E. Rappaport, and V. Tallapragada. 2020. "Recent progress in tropical cyclone intensity forecasting at the National Hurricane Center." *Weather and Forecasting* 35(5):1913-1922. https://doi.org/10.1175/Waf-D-20-0059.1.

Caplan-Auerbach, J., R. P. Dziak, J. Haxel, D. R. Bohnenstiehl, and C. Garcia. 2017. "Explosive processes during the 2015 eruption of Axial Seamount, as recorded by seafloor hydrophones." *Geochemistry Geophysics Geosystems* 18(4):1761-1774. https://doi.org/10.1002/2016gc006734.

Carlton, J. 2024. "Cold weather businesses suffer in the winter that wasn't." *The Wall Street Journal*, March 8. https://www.wsj.com/us-news/climate-environment/cold-weather-businesses-suffer-in-the-winter-that-wasnt-a59ac42b?mod=climate-environment_more_article_pos76 (accessed March 20, 2025).

Carranza, M. M., M. C. Long, A. Di Luca, A. J. Fassbender, K. S. Johnson, Y. Takeshita, P. Mongwe, and K. E. Turner. 2024. "Extratropical storms induce carbon outgassing over the Southern Ocean." *NPJ Climate and Atmospheric Science* 7(1):106. https://doi.org/10.1038/s41612-024-00657-7.

Carter, B. R., R. A. Feely, S. Mecking, J. N. Cross, A. M. Macdonald, S. A. Siedlecki, L. D. Talley, C. L. Sabine, F. J. Millero, J. H. Swift, A. G. Dickson, and K. B. Rodgers. 2017. "Two decades of Pacific anthropogenic carbon storage and ocean acidification along Global Ocean Ship-based Hydrographic Investigations Program sections P16 and P02." *Global Biogeochemical Cycles* 31(2):306-327. https://doi.org/10.1002/2016gb005485.

Cascadia CoPes Hub. 2025. *Welcome to the Cascadia CoPes Hub.* https://cascadiacopeshub.org (accessed September 26, 2024).

Cavaleri Gerhardinger, L., G. G. Moreira Moura, E. McKinley, P. Christie, N. E. Narchi, A. C. Colonese, C. M. Beitl, N. Puniwai, C. C. de Oliveira, I. da Silveira, D. Harry, and M. M. Early Capistrán. 2024. "A call for a cultural shift in oceanography." *Coastal Management* 52(4-5):175-187. https://doi.org/10.1080/089207 53.2024.2370214.

CDC (U.S. Centers for Disease Control and Prevention). 2024. *Harmful Algal Blooms and Your Health.* https://www.cdc.gov/harmful-algal-blooms/about/index.html (accessed March 20, 2025).

Chadwick, W. W., W. S. D. Wilcock, S. L. Nooner, J. W. Beeson, A. M. Sawyer, and T. K. Lau. 2022. "Geodetic monitoring at Axial Seamount since its 2015 eruption reveals a waning magma supply and tightly linked rates of deformation and seismicity." *Geochemistry, Geophysics, Geosystems* 23(1):e2021GC010153. https://doi.org/10.1029/2021gc010153.

Chen, C. Y., S. S. Kahanamoku, A. Tripati, R. A. Alegado, V. R. Morris, K. Andrade, and J. Hosbey. 2022. "Systemic racial disparities in funding rates at the National Science Foundation." *Elife* 11. https://doi.org/10.7554/eLife.83071.

Chen, Z., S. Siedlecki, M. Long, C. M. Petrik, C. A. Stock, and C. A. Deutsch. 2024. "Skillful multiyear prediction of marine habitat shifts jointly constrained by ocean temperature and dissolved oxygen." *Nature Communications* 15(1):900. https://doi.org/10.1038/s41467-024-45016-5.

Cheung, W. W. L., C.-L. Wei, and L. A. Levin. 2022. "Vulnerability of exploited deep-sea demersal species to ocean warming, deoxygenation, and acidification." *Environmental Biology of Fishes* 105(10):1301-1315. https://doi.org/10.1007/s10641-022-01321-w.

Chikamoto, M. O., and P. DiNezio. 2021. "Multi-century changes in the ocean carbon cycle controlled by the tropical oceans and the Southern Ocean." *Global Biogeochemical Cycles* 35(12):e2021GB007090. https://doi.org/10.1029/2021GB007090.

Chisholm, S. W., R. J. Olson, E. R. Zettler, R. Goericke, J. B. Waterbury, and N. A. Welschmeyer. 1988. "A novel free-living prochlorophyte abundant in the oceanic euphotic zone." *Nature* 334(6180):340-343. https://doi.org/10.1038/334340a0.

Christiansen, B., A. Denda, and S. Christiansen. 2020. "Potential effects of deep seabed mining on pelagic and benthopelagic biota." *Marine Policy* 114:103442. https://doi.org/10.1016/j.marpol.2019.02.014.

Christie, P. 2011. "Creating space for interdisciplinary marine and coastal research: Five dilemmas and suggested resolutions." *Environmental Conservation* 38(2):172-186. https://doi.org/10.1017/S0376892911000129.

Ciannelli, L., M. Hunsicker, A. Beaudreau, K. Bailey, L. B. Crowder, C. Finley, C. Webb, J. Reynolds, K. Sagmiller, J. M. Anderies, D. Hawthorne, J. Parrish, S. Heppell, F. Conway, and P. Chigbu. 2014. "Transdisciplinary graduate education in marine resource science and management." *Ices Journal of Marine Science* 71(5):1047-1051. https://doi.org/10.1093/icesjms/fsu067.

Coll, M., and H. K. Lotze. 2016. "Ecological indicators and food-web models as tools to study historical changes in marine ecosystems." In *Perspectives on Oceans Past*, edited by K. Schwerdtner Máñez and B. Poulsen, 103-132. Dordrecht: Springer Netherlands.

Connock, G. T., J. D. Owens, and X.-L. Liu. 2022. "Biotic induction and microbial ecological dynamics of Oceanic Anoxic Event 2." *Communications Earth & Environment* 3(1):136. https://doi.org/10.1038/s43 247-022-00466-x.

Costello, M. J., and C. Chaudhary. 2017. "Marine biodiversity, biogeography, deep-sea gradients, and conservation." *Current Biology* 27(11):R511-R527. https://doi.org/10.1016/j.cub.2017.04.060.

Cowen, R. K., G. Gawarkiewicz, J. Pineda, S. R. Thorrold, and F. E. Werner. 2007. "Population connectivity in marine systems: An overview." *Oceanography* 20(3):14-21. http://www.jstor.org/stable/24860093.

Craig, J. K., and J. S. Link. 2023. "It is past time to use ecosystem models tactically to support ecosystem-based fisheries management: Case studies using Ecopath with Ecosim in an operational management context." *Fish and Fisheries* 24(3):381-406. https://doi.org/10.1111/faf.12733.

Cramer, L. A., C. Flathers, D. Caracciolo, S. M. Russell, and F. Conway. 2018. "Graying of the fleet: Perceived impacts on coastal resilience and local policy." *Marine Policy* 96:27-35. https://doi.org/10.1016/j.marpol. 2018.07.012.

Crichton, K. A., J. D. Wilson, A. Ridgwell, F. Boscolo-Galazzo, E. H. John, B. S. Wade, and P. N. Pearson. 2023. "What the geological past can tell us about the future of the ocean's twilight zone." *Nature Communications* 14(1):2376. https://doi.org/10.1038/s41467-023-37781-6.

Crosman, K. M., E. H. Allison, Y. Ota, A. M. Cisneros-Montemayor, G. G. Singh, W. Swartz, M. Bailey, K. M. Barclay, G. Blume, M. Colléter, M. Fabinyi, E. M. Faustman, R. Fielding, P. J. Griffin, Q. Hanich, H. Harden-Davies, R. P. Kelly, T.-A. Kenny, T. Klinger, J. N. Kittinger, K. Nakamura, A. P. Pauwelussen, S. Pictou, C. Rothschild, K. L. Seto, and A. K. Spalding. 2022. "Social equity is key to sustainable ocean governance." *NPJ Ocean Sustainability* 1(1):4. https://doi.org/10.1038/s44183-022-00001-7.

Crump, B. C., E. V. Armbrust, and J. A. Baross. 1999. "Phylogenetic analysis of particle-attached and free-living bacterial communities in the Columbia River, its estuary, and the adjacent coastal ocean." *Applied and Environmental Microbiology* 65(7):3192-3204. https://doi.org/doi:10.1128/AEM.65.7.3192-3204.1999.

Crump, S. E. 2021. "Sedimentary ancient DNA as a tool in paleoecology." *Nature Reviews Earth & Environment* 2(4):229-229. https://doi.org/10.1038/s43017-021-00158-8.

Curtis, P. E., and A. V. Fedorov. 2024. "Collapse and slow recovery of the Atlantic Meridional Overturning Circulation (AMOC) under abrupt greenhouse gas forcing." *Climate Dynamics*. https://doi.org/10.1007/s00382-024-07185-3.

Dalton, K., M. Skrobe, H. Bell, B. Kantner, D. Berndtson, L. C. Gerhardinger, and P. Christie. 2020. "Marine-related learning networks: Shifting the paradigm toward collaborative ocean governance." *Frontiers in Marine Science* 7. https://doi.org/10.3389/fmars.2020.595054.

Dangendorf, S., Q. Sun, T. Wahl, P. Thompson, J. X. Mitrovica, and B. Hamlington. 2024. "Probabilistic reconstruction of sea-level changes and their causes since 1900." *Earth System Science Data* 16(7):3471-3494. https://doi.org/10.5194/essd-16-3471-2024.

Davidson, K., D. M. Anderson, M. Mateus, B. Reguera, J. Silke, M. Sourisseau, and J. Maguire. 2016. "Forecasting the risk of harmful algal blooms." *Harmful Algae* 53:1-7. https://doi.org/10.1016/j.hal.2015.11.005.

Department of Defense (DOD). n.d. *Spotlight: Tackling the Climate Crisis*. https://www.defense.gov/spotlights/tackling-the-climate-crisis.

DOD. 2021. *Department of Defense Climate Risk Analysis*. Report submitted to the National Security Council Office of the Undersecretary for Policy (Strategy, Plans, and Capabilities).

De Vos, A. 2020. "The problem of 'colonial science.'" *Scientific American* 1(07):2020.

De Vos, A., S. Cambronero-Solano, S. Mangubhai, L. Nefdt, L. C. Woodall, and P. V. Stefanoudis. 2023. "Towards equity and justice in ocean sciences." *NPJ Ocean Sustainability* 2(1):25. https://doi.org/10.1038/s44183-023-00028-4.

De Vos, A., and M. W. Schwartz. 2022. "Confronting parachute science in conservation." *Conservation Science and Practice* 4(5):e12681.

DeFilippo, L. B., L. C. McManus, D. E. Schindler, M. L. Pinsky, M. A. Colton, H. E. Fox, E. W. Tekwa, S. R. Palumbi, T. E. Essington, and M. M. Webster. 2022. "Assessing the potential for demographic restoration and assisted evolution to build climate resilience in coral reefs." *Ecological Applications* 32(7):e2650. https://doi.org/10.1002/eap.2650.

Dixon, O., and A. Gallagher. 2023. "Blue carbon ecosystems and shark behaviour: An overview of key relationships, network interactions, climate impacts, and future research needs." *Frontiers in Marine Science* 10. https://doi.org/10.3389/fmars.2023.1202972.

Dong, J., B. Fox-Kemper, J. O. Wenegrat, A. S. Bodner, X. Yu, S. Belcher, and C. Dong. 2024. "Submesoscales are a significant turbulence source in global ocean surface boundary layer." *Nature Communications* 15(1):9566. https://doi.org/10.1038/s41467-024-53959-y.

Drazen, J. C., C. R. Smith, K. M. Gjerde, S. H. D. Haddock, G. S. Carter, C. A. Choy, M. R. Clark, P. Dutrieux, E. Goetze, C. Hauton, M. Hatta, J. A. Koslow, A. B. Leitner, A. Pacini, J. N. Perelman, T. Peacock, T. T. Sutton, L. Watling, and H. Yamamoto. 2020. "Midwater ecosystems must be considered when evaluating environmental risks of deep-sea mining." *Proceedings of the National Academy of Sciences* 117(30):17455-17460. https://doi.org/doi:10.1073/pnas.2011914117.

Dubilier, N., M. McFall-Ngai, and L. Zhao. 2015. "Microbiology: Create a global microbiome effort." *Nature* 526(7575):631-634. https://doi.org/10.1038/526631a.

Duffy, J. E., J. S. Lefcheck, R. D. Stuart-Smith, S. A. Navarrete, and G. J. Edgar. 2016. "Biodiversity enhances reef fish biomass and resistance to climate change." *Proceedings of the National Academy of Sciences* 113(22):6230-6235. https://doi.org/10.1073/pnas.1524465113.

Dushaw, B. D., P. F. Worcester, W. H. Munk, R. C. Spindel, J. A. Mercer, B. M. Howe, K. Metzger Jr., T. G. Birdsall, R. K. Andrew, M. A. Dzieciuch, B. D. Cornuelle, and D. Menemenlis. 2009. "A decade of acoustic thermometry in the North Pacific Ocean." *Journal of Geophysical Research: Oceans* 114(C7). https://doi.org/10.1029/2008JC005124.

EPA (U.S. Environmental Protection Agency). 2024. *Harmful Algal Blooms Research.* https://www.epa.gov/water-research/harmful-algal-blooms-research#toxicity (accessed March 20, 2025).

Erlandson, J. M. 2001. "The archaeology of aquatic adaptations: Paradigms for a new millennium." *Journal of Archaeological Research* 9(4):287-350. https://doi.org/10.1023/a:1013062712695.

Eskandari, A., T. C. Leow, M. B. A. Rahman, and S. N. Oslan. 2024. "Molecular dynamics-guided insight into the adsorption–inhibition mechanism for controlling ice growth/melt of antifreeze protein type IV mutant from longhorn sculpin fish." *Chemical Papers* 78(7):4437-4454. https://doi.org/10.1007/s11696-024-03407-4.

Feely, R. A., V. J. Fabry, A. Dickson, J.-P. Gattuso, J. Bijma, U. Riebesell, S. Doney, C. Turley, T. Saino, and K. Lee. 2010. "An international observational network for ocean acidification." *Proceedings of Ocean Obs* 9:21-25.

Firth, L. B., A. M. Knights, D. Bridger, A. J. Evans, N. Mieszkowska, P. J. Moore, N. E. O'Connor, E. V. Sheehan, R. C. Thompson, and S. J. Hawkins. 2016. "Ocean sprawl: Challenges and opportunities for biodiversity management in a changing world." *Oceanography and Marine Biology: An Annual Review* 54:189-262.

Fischer, J., J. Jorgensen, H. Josupeit, D. Kalikoski, and C. M. Lucas. 2015. "Fishers' knowledge and the ecosystem approach to fisheries: Applications, experiences and lessons in Latin America." *FAO Fisheries and Aquaculture Technical Paper* (591):I.

Fontela, M., M. I. García-Ibáñez, D. A. Hansell, H. Mercier, and F. F. Pérez. 2016. "Dissolved organic carbon in the North Atlantic meridional overturning circulation." *Scientific Reports* 6(1):26931. https://doi.org/10.1038/srep26931.

Freeman, R. B., and W. Huang. 2014. "Collaboration: Strength in diversity." *Nature* 513(7518):305. https://doi.org/10.1038/513305a.

Fulton, E. A., J. L. Blanchard, J. Melbourne-Thomas, É. E. Plagányi, and V. J. D. Tulloch. 2019. "Where the ecological gaps remain: A modelers' perspective." *Frontiers in Ecology and Evolution* 7. https://doi.org/10.3389/fevo.2019.00424.

Fulton, E. A., J. S. Link, I. C. Kaplan, M. Savina-Rolland, P. Johnson, C. Ainsworth, P. Horne, R. Gorton, R. J. Gamble, A. D. M. Smith, and D. C. Smith. 2011. "Lessons in modelling and management of marine ecosystems: The Atlantis experience." *Fish and Fisheries* 12(2):171-188. https://doi.org/10.1111/j.1467-2979.2011.00412.x.

Gaffney, A. 2024. "It's been the hottest summer on record, European officials say." *The New York Times*, September 5. https://www.nytimes.com/2024/09/05/climate/2024-hottest-summer-on-record.html (accessed March 20, 2025).

Galand, P. E., O. Pereira, C. Hochart, J. C. Auguet, and D. Debroas. 2018. "A strong link between marine microbial community composition and function challenges the idea of functional redundancy." *The ISME Journal* 12(10):2470-2478. https://doi.org/10.1038/s41396-018-0158-1.

Galaz García, C., K. J. Bagstad, J. Brun, R. Chaplin-Kramer, T. Dhu, N. J. Murray, C. J. Nolan, T. H. Ricketts, H. M. Sosik, D. Sousa, G. Willard, and B. S. Halpern. 2023. "The future of ecosystem assessments is automation, collaboration, and artificial intelligence." *Environmental Research Letters* 18(1):011003. https://doi.org/10.1088/1748-9326/acab19.

Gallego, M. A., A. Timmermann, T. Friedrich, and R. E. Zeebe. 2020. "Anthropogenic intensification of surface ocean interannual pCO_2 variability." *Geophysical Research Letters* 47(13):e2020GL087104. https://doi.org/10.1029/2020GL087104.

Gamfeldt, L., J. S. Lefcheck, J. E. K. Byrnes, B. J. Cardinale, J. E. Duffy, and J. N. Griffin. 2015. "Marine biodiversity and ecosystem functioning: what's known and what's next?" *Oikos* 124(3):252-265. https://doi.org/10.1111/oik.01549.

García-Quijano, C. G., and M. V. Pizzini. 2015. "Ecosystem-based knowledge and reasoning in tropical, multispecies, small-scale fishers' LEK: What can fishers LEK contribute to coastal ecological science and management?" *FAO Fisheries & Aquaculture Technical Paper* 591. Rome: Food and Agriculture Organization of the United Nations.

Garcia-Soto, C., L. J. Cheng, L. Caesar, S. Schmidtko, E. B. Jewett, A. Cheripka, I. Rigor, A. Caballero, S. Chiba, J. C. Báez, T. Zielinski, and J. P. Abraham. 2021. "An overview of ocean climate change indicators: Sea surface temperature, ocean heat content, ocean pH, Dissolved oxygen concentration, arctic sea ice extent, thickness and volume, sea level and strength of the AMOC (Atlantic meridional overturning circulation)." *Frontiers in Marine Science* 8. https://doi.org/ARTN64237210.3389/fmars.2021.642372.

Garibay-Toussaint, I., C. Olguín-Jacobson, C. B. Woodson, N. Arafeh-Dalmau, J. Torre, S. Fulton, F. Micheli, R. O'Connor, M. Précoma-de la Mora, A. Hernández-Velasco, and N. E. Narchi. 2024. "Combining the uncombinable: Corporate memories, ethnobiological observations, oceanographic and ecological data to enhance climatic resilience in small-scale fisheries." *Frontiers in Marine Science* 11. https://doi.org/10.3389/fmars.2024.1458059.

Garland, D., F. Lucy, and N. Touzet. 2023. "Summer dynamics of cyanobacteria in an oligo-mesotrophic temperate lake in Northwest Ireland." *Hydrobiologia* 850(19):4327-4341. https://doi.org/10.1007/s10750-023-05307-2.

Gerhardinger, L. C., A. C. Colonese, R. G. Martini, I. da Silveira, A. Zivian, D. F. Herbst, B. Glavovic, S. T. Calvo, and P. Christie. 2024. "Networked media and information ocean literacy: A transformative approach for UN ocean decade." *NPJ Ocean Sustainability* 3(1):2. https://doi.org/10.1038/s44183-023-00038-2.

Germond-Duret, C., C. P. Heidkamp, and J. Morrissey. 2023. "(In)justice and the blue economy." *The Geographical Journal* 189(2):184-192. https://doi.org/10.1111/geoj.12483.

Gilbert P. M., Mitra A. (2022). From webs, loops, shunts, and pumps to microbial multitasking: Evolving concepts of marine microbial ecology, the Mixoplankton Paradigm, and implications for a future ocean. *Limnology and Oceanography* 67:585-597. https://doi.org/10.1002/lno.12018.

Gobler, C. J., and H. Baumann. 2016. "Hypoxia and acidification in ocean ecosystems: Coupled dynamics and effects on marine life." *Biology Letters* 12(5):20150976.

Graham, J., G. Hodsdon, A. Busse, and M. P. Crosby. 2023. "BIPOC voices in ocean sciences: A qualitative exploration of factors impacting career retention." *Journal of Geoscience Education* 71(3):369-387. https://doi.org/10.1080/10899995.2022.2052553.

Graupe, C., L. J. Van Uffelen, P. F. Worcester, M. A. Dzieciuch, and B. M. Howe. 2022. "An automated framework for long-range acoustic positioning of autonomous underwater vehicles." *The Journal of the Acoustical Society of America* 152(3):1615-1626. https://doi.org/10.1121/10.0013830.

Graw, J. H., W. T. Wood, and B. J. Phrampus. 2021. "Predicting global marine sediment density using the random forest regressor machine learning algorithm." *Journal of Geophysical Research: Solid Earth* 126(1):e2020JB020135. https://doi.org/10.1029/2020JB020135.

Gray, A. R. 2024. "The four-dimensional carbon cycle of the Southern Ocean." *Annual Review of Marine Science* 16:163-190. https://doi.org/10.1146/annurev-marine-041923-104057.

Gray, B. 2008. "Enhancing transdisciplinary research through collaborative leadership." *American Journal of Preventive Medicine* 35(2 Suppl):S124-32. https://doi.org/10.1016/j.amepre.2008.03.037.

Greene, C. H., C. M. Scott-Buechler, A. L. P. Hausner, Z. I. Johnson, X. G. Lei, and M. E. Huntley. 2022. "Transforming the future of marine aquaculture: A circular economy approach." *Oceanography* 35:26-34. https://doi.org/10.5670/oceanog.2022.213.

Grégoire, M., A. Oschlies, D. Canfield, C. Castro, I. Ciglenecki, P. Croot, K. Salin, B. Schneider, P. Serret, and C. Slomp. 2023. *Ocean Oxygen: The Role of the Ocean in the Oxygen We Breathe and the Threat of Deoxygenation.* Ostend, Belgium: European Marine Board.

Gregor, L., and N. Gruber. 2021. "OceanSODA-ETHZ: A global gridded data set of the surface ocean carbonate system for seasonal to decadal studies of ocean acidification." *Earth System Science Data* 13(2):777-808. https://doi.org/10.5194/essd-13-777-2021.

Gruber, N., D. C. E. Bakker, T. DeVries, L. Gregor, J. Hauck, P. Landschuetzer, G. A. McKinley, and J. D. Mueller. 2023. "Trends and variability in the ocean carbon sink." *Nature Reviews Earth & Environment* 4(2):119-134. https://doi.org/10.1038/s43017-022-00381-x.

Gruber, N., P. Landschutzer, and N. S. Lovenduski. 2019. "The variable Southern Ocean carbon sink." *Annual Review of Marine Science* 11:159-186. https://doi.org/10.1146/annurev-marine-121916-063407.

Gulev, S. K., P. W. Thorne, J. Ahn, F. J. Dentener, C. M. Domingues, S. Gerland, D. Gong, D. S. Kaufman, H. C. Nnamchi, and J. Quaas. 2021. "Changing state of the climate system." In the *Sixth Assessment Report of the Intergovernmental Panel on Climate Change.* Cambridge: Cambridge University Press.

Gumusay, M. U., T. Bakirman, I. Tuney Kizilkaya, and N. O. Aykut. 2019. "A review of seagrass detection, mapping and monitoring applications using acoustic systems." *European Journal of Remote Sensing* 52(1):1-29. https://doi.org/10.1080/22797254.2018.1544838.

Hall, K. L., A. L. Vogel, B. Stipelman, D. Stokols, G. Morgan, and S. Gehlert. 2012. "A four-phase model of transdisciplinary team-based research: Goals, team processes, and strategies." *Translational Behavioral Medicine* 2(4):415-430. https://doi.org/10.1007/s13142-012-0167-y.

Hall, T. M., and J. P. Kossin. 2019. "Hurricane stalling along the North American coast and implications for rainfall." *NPJ Climate and Atmospheric Science* 2(1):17. https://doi.org/10.1038/s41612-019-0074-8.

Hamdan, L. J., R. B. Coffin, M. Sikaroodi, J. Greinert, T. Treude, and P. M. Gillevet. 2012. "Ocean currents shape the microbiome of Arctic marine sediments." *The ISME Journal* 7(4):685-696. https://doi.org/10.1038/ismej.2012.143.

Hamlington, B. D., A. Bellas-Manley, J. K. Willis, S. Fournier, N. Vinogradova, R. S. Nerem, C. G. Piecuch, P. R. Thompson, and R. Kopp. 2024. "The rate of global sea level rise doubled during the past three decades." *Communications Earth & Environment* 5(1):601. https://doi.org/10.1038/s43247-024-01761-5.

Hamlington, B. D., A. S. Gardner, E. Ivins, J. T. M. Lenaerts, J. T. Reager, D. S. Trossman, E. D. Zaron, S. Adhikari, A. Arendt, A. Aschwanden, B. D. Beckley, D. P. S. Bekaert, G. Blewitt, L. Caron, D. P. Chambers, H. A. Chandanpurkar, K. Christianson, B. Csatho, R. I. Cullather, R. M. DeConto, J. T. Fasullo, T. Frederikse, J. T. Freymueller, D. M. Gilford, M. Girotto, W. C. Hammond, R. Hock, N. Holschuh, R. E. Kopp, F. Landerer, E. Larour, D. Menemenlis, M. Merrifield, J. X. Mitrovica, R. S. Nerem, I. J. Nias, V. Nieves, S. Nowicki, K. Pangaluru, C. G. Piecuch, R. D. Ray, D. R. Rounce, N.-J. Schlegel, H. Seroussi, M. Shirzaei, W. V. Sweet, I. Velicogna, N. Vinogradova, T. Wahl, D. N. Wiese, and M. J. Willis. 2020. "Understanding of contemporary regional sea-level change and the implications for the future." *Reviews of Geophysics* 58(3):e2019RG000672. https://doi.org/10.1029/2019RG000672.

Hampel, J. J., R. D. Moseley, and L. J. Hamdan. 2023. "Microbiomes respond predictably to built habitats on the seafloor." *Molecular Ecology* 32(23):6686-6695. https://doi.org/10.1111/mec.16504.

Han, Z. Y., and H. O. Sharif. 2021. "Analysis of flood fatalities in the United States, 1959-2019." *Water* 13(13):1871. https://doi.org/ARTN 187110.3390/w13131871.

Harðardóttir, S., J. S. Haile, J. L. Ray, A. Limoges, N. Van Nieuwenhove, C. Lalande, P.-L. Grondin, R. Jackson, K. S. Skaar, M. Heikkilä, J. Berge, N. Lundholm, G. Massé, S. Rysgaard, M.-S. Seidenkrantz, S. De Schepper, E. D. Lorenzen, C. Lovejoy, and S. Ribeiro. 2024. "Millennial-scale variations in Arctic sea ice are recorded in sedimentary ancient DNA of the microalga Polarella glacialis." *Communications Earth & Environment* 5(1):25. https://doi.org/10.1038/s43247-023-01179-5.

Harris, L. A., T. Grayson, H. A. Neckles, C. T. Emrich, K. A. Lewis, K. W. Grimes, S. Williamson, C. Garza, C. R. Whitcraft, J. B. Pollack, D. M. Talley, B. Fertig, C. M. Palinkas, S. Park, J. M. P. Vaudrey, A. M. Fitzgerald, and J. Quispe. 2022. "A socio-ecological imperative for broadening participation in coastal and estuarine research and management." *Estuaries and Coasts* 45(1):38-48. https://doi.org/10.1007/s12237-021-00944-z.

Hauck, J., L. Gregor, C. Nissen, L. Patara, M. Hague, P. Mongwe, S. Bushinsky, S. C. Doney, N. Gruber, C. Le Quéré, M. Manizza, M. Mazloff, P. M. S. Monteiro, and J. Terhaar. 2023. "The Southern Ocean Carbon Cycle 1985–2018: Mean, seasonal cycle, trends, and storage." *Global Biogeochemical Cycles* 37(11):e2023GB007848. https://doi.org/10.1029/2023GB007848.

Haugen, B. I., L. A. Cramer, G. G. Waldbusser, and F. D. L. Conway. 2021. "Resilience and adaptive capacity of Oregon's fishing community: Cumulative impacts of climate change and the graying of the fleet." *Marine Policy* 126:104424. https://doi.org/10.1016/j.marpol.2021.104424.

Heery, E. C., M. J. Bishop, L. P. Critchley, A. B. Bugnot, L. Airoldi, M. Mayer-Pinto, E. V. Sheehan, R. A. Coleman, L. H. L. Loke, E. L. Johnston, V. Komyakova, R. L. Morris, E. M. A. Strain, L. A. Naylor, and K. A. Dafforn. 2017. "Identifying the consequences of ocean sprawl for sedimentary habitats." *Journal of Experimental Marine Biology and Ecology* 492:31-48. https://doi.org/10.1016/j.jembe.2017.01.020.

Heinze, C., T. Blenckner, H. Martins, D. Rusiecka, R. Döscher, M. Gehlen, N. Gruber, E. Holland, Ø. Hov, F. Joos, J. B. R. Matthews, R. Rødven, and S. Wilson. 2021. "The quiet crossing of ocean tipping points." *Proceedings of the National Academy of Sciences* 118(9):e2008478118. https://doi.org/ 10.1073/pnas. 2008478118.

Hendricks, A., C. M. Mackie, E. Luy, C. Sonnichsen, J. Smith, I. Grundke, M. Tavasoli, A. Furlong, R. G. Beiko, J. LaRoche, and V. Sieben. 2023. "Compact and automated eDNA sampler for in situ monitoring of marine environments." *Scientific Reports* 13(1):5210. https://doi.org/10.1038/s41598-023-32310-3.

Henson, S. A., C. Laufkötter, S. Leung, S. L. C. Giering, H. I. Palevsky, and E. L. Cavan. 2022. "Uncertain response of ocean biological carbon export in a changing world." *Nature Geoscience* 15(4):248-254. https://doi.org/10.1038/s41561-022-00927-0.

Heymans, J. J., M. Coll, S. Libralato, L. Morissette, and V. Christensen. 2014. "Global patterns in ecological indicators of marine food webs: A modelling approach." *PLoS One* 9(4):e95845. https://doi.org/10.1371/ journal.pone.0095845.

Heymans, J. J., M. Coll, J. S. Link, S. Mackinson, J. Steenbeek, C. Walters, and V. Christensen. 2016. "Best practice in Ecopath with Ecosim food-web models for ecosystem-based management." *Ecological Modelling* 331:173-184. https://doi.org/10.1016/j.ecolmodel.2015.12.007.

Hills, J. M., and P. N. Maharaj. 2023. "Designing transdisciplinarity for transformative ocean governance." *Frontiers in Marine Science* 10. https://doi.org/ARTN 1075759.

Hoehler, T. M., D. J. Mankel, P. R. Girguis, T. M. McCollom, N. Y. Kiang, and B. B. Jørgensen. 2023. "The metabolic rate of the biosphere and its components." *Proceedings of the National Academy of Sciences* 120(25):e2303764120. https://doi.org/10.1073/pnas.2303764120.

Hoelting, K., B. Moore, R. Pollnac, and P. Christie. 2014. "Collaboration within the Puget Sound Marine and Nearshore Science Network." *Coastal Management* 42(4):332-354. https://doi.org/10.1080/08920753.20 14.923141.

Hoffmann, S., L. Deutsch, J. T. Klein, and M. O'Rourke. 2022. "Integrate the integrators! A call for establishing academic careers for integration experts." *Humanities and Social Sciences Communications* 9(1):147. https://doi.org/10.1057/s41599-022-01138-z.

Holling. 1973. "C. S. Holling (1973)." In *Foundations of Socio-Environmental Research: Legacy Readings with Commentaries*, edited by William R. Burnside, Simone Pulver, Kathryn J. Fiorella, Meghan L. Avolio and Steven M. Alexander, pp. 460-482. Cambridge: Cambridge University Press.

Holt, C. C., E. Hehenberger, D. V. Tikhonenkov, V. K. L. Jacko-Reynolds, N. Okamoto, E. C. Cooney, N. A. T. Irwin, and P. J. Keeling. 2023. "Multiple parallel origins of parasitic Marine Alveolates." *Nature Communications* 14(1):7049. https://doi.org/10.1038/s41467-023-42807-0.

Honisch, B., D. L. Royer, D. O. Breecker, P. J. Polissar, G. J. Bowen, M. J. Henehan, Y. Cui, M. Steinthorsdottir, J. C. McElwain, M. J. Kohn, A. Pearson, S. R. Phelps, K. T. Uno, A. Ridgwell, E. Anagnostou, J. Austermann, M. P. S. Badger, R. S. Barclay, P. K. Bijl, T. B. Chalk, C. R. Scotese, E. de la Vega, R. M. DeConto, K. A. Dyez, V. Ferrini, P. J. Franks, C. F. Giulivi, M. Gutjahr, D. T. Harper, L. L. Haynes, M. Huber, K. E. Snell, B. A. Keisling, W. Konrad, T. K. Lowenstein, A. Malinverno, M. Guillermic, L. M. Mejia, J. N. Milligan, J. J. Morton, L. Nordt, R. Whiteford, A. Roth-Nebelsick, J. K. C. Rugenstein, M. F. Schaller, N. D. Sheldon, S. Sosdian, E. B. Wilkes, C. R. Witkowski, Y. G. Zhang, L. Anderson, D. J. Beerling, C. Bolton, T. E. Cerling, J. M. Cotton, J. Da, D. D. Ekart, G. L. Foster, D. R. Greenwood, E. G. Hyland, E. A. Jagniecki, J. P. Jasper, J. B. Kowalczyk, L. Kunzmann, W. M. Kurschner, C. E. Lawrence, C. H. Lear, M. A. Martinez-Boti, D. P. Maxbauer, P. Montagna, B. D. A. Naafs, J. W. B. Rae, M. Raitzsch, G. J. Retallack, S. J. Ring, O. Seki, J. Sepulveda, A. Sinha, T. F. Tesfamichael, A. Tripati, J. van der Burgh, J. Yu, J. C. Zachos, L. Zhang, and Cenozoic C. O. Proxy Integration Project Consortium. 2023. "Toward a Cenozoic history of atmospheric CO$_2$." *Science* 382(6675):eadi5177. https://doi.org/10.1126/science. adi5177.

Horn, A., M. W. Visser, C. A. C. M. Pittens, E. Urias, M. B. M. Zweekhorst, and G. M. van Dijk. 2024. "Transdisciplinary learning trajectories: Developing action and attitude in interplay." *Humanities and Social Sciences Communications* 11(1):149. https://doi.org/10.1057/s41599-023-02541-w.

Howe, B. M., B. K. Arbic, J. Aucan, C. R. Barnes, N. Bayliff, N. Becker, R. Butler, L. Doyle, S. Elipot, G. C. Johnson, F. Landerer, S. Lentz, D. S. Luther, M. Müller, J. Mariano, K. Panayotou, C. Rowe, H. Ota, Y. T. Song, M. Thomas, P. N. Thomas, P. Thompson, F. Tilmann, T. Weber, and S. Weinstein. 2019.

"SMART cables for observing the global ocean: Science and implementation." *Frontiers in Marine Science* 6. https://doi.org/10.3389/fmars.2019.00424.

International Ocean Discovery Program (IODP). 2024. *JOIDES Resolution Facility Board Meeting.* (Honolulu, Hawai'i). https://www.iodp.org/jrfb-minutes/1233-jrfb-2024-may-minutes/file.

International Polar Year. n.d. *Why an International Polar Year (IPY) in 2032–33?* https://www.ipy5.info/about/why-an-international-polar-year-ipy-in-2032-33.

IPBES (Intergovernmental Science-Policy Platform on Biodiversity and Ecosystem Services). 2019. *The Global Assessment Report on Biodiversity and Ecosystem Services: Summary for Policy Makers.* Bonn, Germany: IPBES.

IPCC (Intergovernmental Panel on Climate Change). 2021. *Climate Change 2021—The Physical Science Basis: Working Group I Contribution to the Sixth Assessment Report of the Intergovernmental Panel on Climate Change.* Cambridge: Cambridge University Press.

IPCC. 2023. *Climate Change 2023: Synthesis Report. Contribution of Working Groups I, II and III to the Sixth Assessment Report of the Intergovernmental Panel on Climate Change*, edited by H. Lee and J. Romero, pp. 35-115. https://doi.org/10.59327/IPCC/AR6-9789291691647.

ISTC (Subcommittee on International Science & Technology Coordination). 2024. *Biennial Report to Congress on International Science & Technology Cooperation.* Washington, DC: National Science and Technology Council.

Ito, Y., R. Hino, M. Kido, H. Fujimoto, Y. Osada, D. Inazu, Y. Ohta, T. Iinuma, M. Ohzono, S. Miura, M. Mishina, K. Suzuki, T. Tsuji, and J. Ashi. 2013. "Episodic slow slip events in the Japan subduction zone before the 2011 Tohoku-Oki earthquake." *Tectonophysics* 600:14-26. https://doi.org/10.1016/j.tecto.2012.08.022.

IUCN (International Union for the Conservation of Nature). 2012. *IUCN Red List Categories and Criteria. Version 3.1.* Gland, Switzerland: IUCN.

Ivakhiv, A. 2002. "Toward a multicultural ecology." *Organization & Environment* 15(4):389-409.

Jackson, J. B., M. X. Kirby, W. H. Berger, K. A. Bjorndal, L. W. Botsford, B. J. Bourque, R. H. Bradbury, R. Cooke, J. Erlandson, J. A. Estes, T. P. Hughes, S. Kidwell, C. B. Lange, H. S. Lenihan, J. M. Pandolfi, C. H. Peterson, R. S. Steneck, M. J. Tegner, and R. R. Warner. 2001. "Historical overfishing and the recent collapse of coastal ecosystems." *Science* 293(5530):629-37. https://doi.org/10.1126/science.1059199.

Jansson, E., E. Faust, D. Bekkevold, M. Quintela, C. Durif, K. T. Halvorsen, G. Dahle, C. Pampoulie, J. Kennedy, B. Whittaker, L. Unneland, S. Post, C. André, and K. A. Glover. 2023. "Global, regional, and cryptic population structure in a high gene-flow transatlantic fish." *PLoS ONE* 18(3):e0283351. https://doi.org/10.1371/journal.pone.0283351.

Jaureguiberry, P., N. Titeux, M. Wiemers, D. E. Bowler, L. Coscieme, A. S. Golden, C. A. Guerra, U. Jacob, Y. Takahashi, J. Settele, S. Díaz, Z. Molnár, and A. Purvis. 2022. "The direct drivers of recent global anthropogenic biodiversity loss." *Science Advances* 8(45):eabm9982. https://doi.org/10.1126/sciadv.abm9982.

Jenkins, S., C. Smith, M. Allen, and R. Grainger. 2023. "Tonga eruption increases chance of temporary surface temperature anomaly above 1.5 °C." *Nature Climate Change* 13(2):127-129. https://doi.org/10.1038/s41558-022-01568-2.

Jobstvogt, N., M. Townsend, U. Witte, and N. Hanley. 2014. "How can we identify and communicate the ecological value of deep-sea ecosystem services?" *PLoS ONE* 9(7):e100646. https://doi.org/10.1371/journal.pone.0100646.

Jodoin, S., A. Savaresi, and M. Wewerinke-Singh. 2021. "Rights-based approaches to climate decision-making." *Current Opinion in Environmental Sustainability* 52:45-53. https://doi.org/10.1016/j.cosust.2021.06.004.

Johnson, G. C., S. Hosoda, S. R. Jayne, P. R. Oke, S. C. Riser, D. Roemmich, T. Suga, V. Thierry, S. E. Wijffels, and J. Xu. 2022. "Argo—Two decades: Global oceanography, revolutionized." *Annual Review of Marine Science* 14:379-403. https://doi.org/10.1146/annurev-marine-022521-102008.

Johnson, T. R., and M. D. Mazur. 2018. "A mixed method approach to understanding the graying of Maine's lobster fleet." *Bulletin of Marine Science* 94(3):1185-1199. https://doi.org/10.5343/bms.2017.1108.

Jørgensen, B. B., and I. P. Marshall. 2016. "Slow microbial life in the seabed." *Annual Review of Marine Science* 8:311-32. https://doi.org/10.1146/annurev-marine-010814-015535.

Kaikkonen, L., R. J. Shellock, S. A. Selim, R. A. Ojwala, B. S. Dias, S. Li, C. I. Addey, I. Gianelli, K. M. Maltby, S. Garcia-Morales, J. Palacios-Abrantes, S. Jiang, M. Albo-Puigserver, V. A. García Alonso, C. A. Baker, C. B. Bove, S. Brodie, L. I. Dahlet, J. Das, A. Dunne, S. C. A. Ferse, E. Johannesen, J. Jung, E. Merayo Garcia, D. B. Karcher, S. Mahadeo, L. Millan, K. O. Lawal, A. Oloko, K. Ortega-Cisneros, S. Otoabasi-Akpan, D. Roy, S. S. Rouf, S. Smoliński, N. Vaidianu, C. Whidden, and M. Strand. 2024. "Fostering diversity, equity, and inclusion in interdisciplinary marine science." *NPJ Ocean Sustainability* 3(1):49. https://doi.org/10.1038/s44183-024-00087-1.

Kappel, E. S., B. E. Cuker, C. Garza, D. Gibson, C. Martinez, W. F. Todd, and C. Xu. 2023. "Introduction to the special issue on building diversity, equity, and inclusion in the ocean sciences." *Oceanography* 36(4):6-9. https://www.jstor.org/stable/27278246.

Kato, A., K. Obara, T. Igarashi, H. Tsuruoka, S. Nakagawa, and N. Hirata. 2012. "Propagation of slow slip leading up to the 2011 M_w 9.0 Tohoku-Oki earthquake." *Science* 335(6069):705-708. https://doi.org/10.1126/science.1215141.

Kelley, D. S., J. R. Delaney, W. Chadwick, B. T. Philip, and S. G. Merle. 2015. "Axial Seamount 2015 eruption: A 127 m thick, microbially-covered lava flow." *AGU Fall Meeting Abstracts,* San Francisco, California.

Kelley, D. S., J. R. Delaney, and S. K. Juniper. 2014. "Establishing a new era of submarine volcanic observatories: Cabling Axial Seamount and the Endeavour segment of the Juan de Fuca Ridge." *Marine Geology* 352:426-450. https://doi.org/10.1016/j.margeo.2014.03.010.

Kiel Reese, B., M. S. Sobol, M. W. Bowles, and K.-U. Hinrichs. 2021. "Redefining the subsurface biosphere: Characterization of fungi isolated from energy-limited marine deep subsurface sediment." *Frontiers in Fungal Biology* 2. https://doi.org/10.3389/ffunb.2021.727543.

Kiessling, W., D. Dimitrijević, N. B. Raja, K. Frühbeißer, A. Vescogni, and F. R. Bosellini. 2024. "Census-based estimates of Mediterranean Oligocene–Miocene reef carbonate production." *Facies* 71(1):2. https://doi.org/10.1007/s10347-024-00692-z.

Kitolelei, S., A. Breckwoldt, J. Kitolelei, and N. Makhoul. 2022. "Fisherwomen's Indigenous and local knowledge-the hidden gems for the management of marine and freshwater resources in Fiji." *Frontiers in Marine Science* 9. https://doi.org/ARTN 99125310.3389/fmars.2022.991253.

Kliskey, A. A., P. Williams, E. J. Trammell, D. Cronan, D. Griffith, L. Alessa, R. Lammers, M. E. d. Haro-Martí, and J. Oxarango-Ingram. 2023. "Building trust, building futures: Knowledge co-production as relationship, design, and process in transdisciplinary science." *Frontiers in Environmental Science* 11. https://doi.org/10.3389/fenvs.2023.1007105.

Klotzbach, P. J., K. M. Wood, C. J. Schreck III, S. G. Bowen, C. M. Patricola, and M. M. Bell. 2022. "Trends in global tropical cyclone activity: 1990–2021." *Geophysical Research Letters* 49(6):e2021GL095774. https://doi.org/10.1029/2021GL095774.

Kluger, M. O., and G. Bartzke. 2020. "A practical guideline how to tackle interdisciplinarity—A synthesis from a post-graduate group project." *Humanities and Social Sciences Communications* 7(1):47. https://doi.org/10.1057/s41599-020-00540-9.

Kousioras, V., and D. Gatopoulos. 2024. "A climate-related mass die-off leaves over 100 tons of dead fish collecting at a Greek port." *AP News*, August 30. https://apnews.com/article/greece-volos-climate-drought-floods-fish-e924a1f9345f26644d17a62b6fe93dcc (accessed March 20, 2025).

Krumhardt, K. M., M. C. Long, C. M. Petrik, M. Levy, F. S. Castruccio, K. Lindsay, L. Romashkov, A.-L. Deppenmeier, R. Denéchère, Z. Chen, L. Landrum, G. Danabasoglu, and P. Chang. 2024. "From nutrients to fish: Impacts of mesoscale processes in a global CESM-FEISTY eddying ocean model framework." *Progress in Oceanography* 227:103314. https://doi.org/10.1016/j.pocean.2024.103314.

Kruse, G. H. 2023. "Are crabs in hot water?" *Science* 382(6668):260-261. https://doi.org/10.1126/science.adk7565.

Kūlana Noiʻi Working Group. 2021. *Kūlana Noiʻi v. 2.* Honolulu: University of Hawaiʻi Sea Grant College Program.

Lacour, L., J. Llort, N. Briggs, P. G. Strutton, and P. W. Boyd. 2023. "Seasonality of downward carbon export in the Pacific Southern Ocean revealed by multi-year robotic observations." *Nature Communications* 14(1):1278. https://doi.org/10.1038/s41467-023-36954-7.

Laurance, W. F. 2013. "Tipping points and the vulnerability of Australia's tropical ecosystems." In *Conservation Biology*: 167-180. https://10.1002/9781118679838.ch20.

Laurenceau-Cornec, E. C., M. Mongin, T. W. Trull, M. Bressac, E. L. Cavan, L. T. Bach, F. A. C. Le Moigne, F. Planchon, and P. W. Boyd. 2023. "Concepts Toward a global mechanistic mapping of ocean carbon export." *Global Biogeochemical Cycles* 37(9):e2023GB007742. https://doi.org/10.1029/2023GB007742.

Le, J. T., P. R. Girguis, and L. A. Levin. 2022. "Using deep-sea images to examine ecosystem services associated with methane seeps." *Marine Environmental Research* 181:105740. https://doi.org/10.1016/j.marenvres.2022.105740.

Lebreton, L., R. de Vries, Y. Pham, H. Wolter, M. van Vulpen, P. Puskic, B. Sainte-Rose, S.-J. Royer, and M. Egger. 2024. "Seven years into the North Pacific garbage patch: Legacy plastic fragments rising disproportionally faster than larger floating objects." *Environmental Research Letters* 19(12):124054.

Levin, L. A., D. J. Amon, and H. Lily. 2020. "Challenges to the sustainability of deep-seabed mining." *Nature Sustainability* 3 (10):784-794. https://doi.org/10.1038/s41893-020-0558-x.

Levin, S. A., and J. Lubchenco. 2008. "Resilience, robustness, and marine ecosystem-based management." *BioScience* 58(1):27-32. https://doi.org/10.1641/b580107.

Lewis, K. A., R. R. Christian, C. W. Martin, K. L. Allen, A. M. McDonald, V. M. Roberts, M. N. Shaffer, and J. F. Valentine. 2021. "Complexities of disturbance response in a marine food web." *Limnology and Oceanography* 67(S1):S352-S364. https://doi.org/10.1002/lno.11790.

Lewis, S. A., A. Holloway, and K. Yarincik. 2023. "Assessing diversity in US ocean science institutions: Insights from fifteen years (2007-2021) of OSER data." *Oceanography* 36(4):10-20. https://doi.org/10.5670/oceanog.2024.134.

Li, L., R. W. Schmitt, C. C. Ummenhofer, and K. B. Karnauskas. 2016. "Implications of North Atlantic sea surface salinity for summer precipitation over the U.S. Midwest: Mechanisms and predictive value." *Journal of Climate* 29(9):3143-3159. https://doi.org/10.1175/JCLI-D-15-0520.1.

Li, Z., M. H. England, and S. Groeskamp. 2023. "Recent acceleration in global ocean heat accumulation by mode and intermediate waters." *Nature Communications* 14(1):6888. https://doi.org/10.1038/s41467-023-42468-z.

Libralato, S., F. Pranovi, M. Zucchetta, M. Anelli Monti, and J. S. Link. 2019. "Global thresholds in properties emerging from cumulative curves of marine ecosystems." *Ecological Indicators* 103:554-562. https://doi.org/10.1016/j.ecolind.2019.03.053.

Lin, Y., C. B. Frey, and L. Wu. 2023. "Remote collaboration fuses fewer breakthrough ideas." *Nature* 623(7989):987-991. https://doi.org/10.1038/s41586-023-06767-1.

Link, J. S., S. Thur, G. Matlock, and M. Grasso. 2023. "Why we need weather forecast analogues for marine ecosystems." *ICES Journal of Marine Science* 80(8):2087-2098. https://doi.org/10.1093/icesjms/fsad143.

Liu, T., H.-W. Ou, X. Liu, and D. Chen. 2022. "On the role of eddy mixing in the subtropical ocean circulation." *Frontiers in Marine Science* 9. https://doi.org/10.3389/fmars.2022.832992.

Longo, G. C., J. Harms, J. R. Hyde, M. T. Craig, A. Ramón-Laca, and K. M. Nichols. 2022. "Genome-wide markers reveal differentiation between and within the cryptic sister species, sunset and vermilion rockfish." *Conservation Genetics* 23(1):75-89. https://doi.org/10.1007/s10592-021-01397-4.

Lotze, H. K., M. Coll, A. M. Magera, C. Ward-Paige, and L. Airoldi. 2011. "Recovery of marine animal populations and ecosystems." *Trends in Ecology & Evolution* 26(11):595-605. https://doi.org/10.1016/j.tree.2011.07.008.

Louca, S., M. F. Polz, F. Mazel, M. B. N. Albright, J. A. Huber, M. I. O'Connor, M. Ackermann, A. S. Hahn, D. S. Srivastava, S. A. Crowe, M. Doebeli, and L. W. Parfrey. 2018. "Function and functional redundancy in microbial systems." *Nature Ecology & Evolution* 2(6):936-943. https://doi.org/10.1038/s41559-018-0519-1.

Ma, S., D. Liu, Y. Tian, C. Fu, J. Li, P. Ju, P. Sun, Z. Ye, Y. Liu, and Y. Watanabe. 2021. "Critical transitions and ecological resilience of large marine ecosystems in the Northwestern Pacific in response to global warming." *Global Change Biology* 27(20):5310-5328. https://doi.org/10.1111/gcb.15815.

MacGilchrist, G. A., A. C. Naveira Garabato, P. J. Brown, L. Jullion, S. Bacon, D. C. E. Bakker, M. Hoppema, M. P. Meredith, and S. Torres-Valdés. 2019. "Reframing the carbon cycle of the subpolar Southern Ocean." *Science Advances* 5(8):eaav6410. https://doi.org/doi:10.1126/sciadv.aav6410.

MacKinnon, J. A., J. D. Nash, M. H. Alford, A. J. Lucas, J. B. Mickett, E. L. Shroyer, A. F. Waterhouse, A. Tandon, D. Sengupta, A. Mahadevan, M. Ravichandran, R. Pinkel, D. L. Rudnick, C. B. Whalen, M. S. Alberty, J. S. Lekha, E. C. Fine, D. Chaudhuri, and G. L. Wagner. 2016. "A tale of two spicy seas." *Oceanography* 29(2):50-61. https://doi.org/10.5670/oceanog.2016.38.

MacKinnon, J. A., H. L. Simmons, J. Hargrove, J. Thomson, T. Peacock, M. H. Alford, B. I. Barton, S. Boury, S. D. Brenner, N. Couto, S. L. Danielson, E. C. Fine, H. C. Graber, J. Guthrie, J. E. Hopkins, S. R. Jayne, C. Jeon, T. Klenz, C. M. Lee, Y. D. Lenn, A. J. Lucas, B. Lund, C. Mahaffey, L. Norman, L. Rainville, M. M. Smith, L. N. Thomas, S. Torres-Valdes, and K. R. Wood. 2021. "A warm jet in a cold ocean." *Nature Communications* 12(1):2418. https://doi.org/10.1038/s41467-021-22505-5.

Margolskee, A., H. Frenzel, S. Emerson, and C. Deutsch. 2019. "Ventilation pathways for the North Pacific Oxygen Deficient Zone." *Global Biogeochemical Cycles* 33(7):875-890. https://doi.org/10.1029/2018gb006149.

Marín-Spiotta, E., R. T. Barnes, A. A. Berhe, M. G. Hastings, A. Mattheis, B. Schneider, and B. M. Williams. 2020. "Hostile climates are barriers to diversifying the geosciences." *Advances in Geosciences* 53:117-127. https://doi.org/10.5194/adgeo-53-117-2020.

Marschalek, J. W., L. Zurli, F. Talarico, T. van de Flierdt, P. Vermeesch, A. Carter, F. Beny, V. Bout-Roumazeilles, F. Sangiorgi, S. R. Hemming, L. F. Pérez, F. Colleoni, J. G. Prebble, T. E. van Peer, M. Perotti, A. E. Shevenell, I. Browne, D. K. Kulhanek, R. Levy, D. Harwood, N. B. Sullivan, S. R. Meyers, E. M. Griffith, C. D. Hillenbrand, E. Gasson, M. J. Siegert, B. Keisling, K. J. Licht, G. Kuhn, J. P. Dodd, C. Boshuis, L. De Santis, R. M. McKay, J. Ash, F. Beny, I. M. Browne, G. Cortese, L. De Santis, J. P. Dodd, O. M. Esper, J. A. Gales, D. M. Harwood, S. Ishino, B. A. Keisling, S. Kim, S. Kim, D. K. Kulhanek, J. S. Laberg, R. M. Leckie, R. M. McKay, J. Müller, M. O. Patterson, B. W. Romans, O. E. Romero, F. Sangiorgi, O. Seki, A. E. Shevenell, S. M. Singh, I. M. Cordeiro de Sousa, S. T. Sugisaki, T. van de Flierdt, T. E. van Peer, W. Xiao, Z. Xiong, and I. Expedition. 2021. "A large West Antarctic Ice Sheet explains early Neogene sea-level amplitude." *Nature* 600(7889):450-455. https://doi.org/10.1038/s41586-021-04148-0.

Marti, F., B. Meyssignac, V. Rousseau, M. Ablain, R. Fraudeau, A. Blazquez, and S. Fourest. 2024. "Monitoring global ocean heat content from space geodetic observations to estimate the Earth energy imbalance." *Copernicus Ocean State Report, Eighth Edition* 4-osr8:3. https://doi.org/10.5194/sp-4-osr8-3-2024.

Matsumoto, G. I., K. S. Johnson, S. Riser, L. Talley, S. Wijffels, and R. Hotinski. 2022. "The Global Ocean Biogeochemistry (GO-BGC) Array of profiling floats to observe changing ocean chemistry and biology." *Marine Technology Society Journal* 56(3):122-123.

McCabe, R. M., B. M. Hickey, R. M. Kudela, K. A. Lefebvre, N. G. Adams, B. D. Bill, F. M. D. Gulland, R. E. Thomson, W. P. Cochlan, and V. L. Trainer. 2016. "An unprecedented coastwide toxic algal bloom linked to anomalous ocean conditions." *Geophysical Research Letters* 43(19):10,366-10,376. https://doi.org/10.1002/2016GL070023.

McCaig, A., S. Q. Lang, P. Blum, and E. Scientists. 2024. *Expedition 399 Preliminary Report: Building Blocks of Life, Atlantis Massif.* International Ocean Discovery Program. http://publications.iodp.org/preliminary _report/399 (accessed March 20, 2025).

McCarthy, G. D., and L. Caesar. 2023. "Can we trust projections of AMOC weakening based on climate models that cannot reproduce the past?" *Philosophical Transactions of the Royal Society A: Mathematical, Physical and Engineering Sciences* 381(2262):20220193. https://doi.org/10.1098/rsta.2022.0193.

McCauley, D. J., M. L. Pinsky, S. R. Palumbi, J. A. Estes, F. H. Joyce, and R. R. Warner. 2015. "Marine defaunation: Animal loss in the global ocean." *Science* 347(6219):1255641. https://doi.org/10.1126/scien ce.1255641.

McGill, B. J., B. J. Enquist, E. Weiher, and M. Westoby. 2006. "Rebuilding community ecology from functional traits." *Trends in Ecology & Evolution* 21(4):178-85. https://doi.org/10.1016/j.tree.2006.02.002.

McManus, L. C., D. L. Forrest, E. W. Tekwa, D. E. Schindler, M. A. Colton, M. M. Webster, T. E. Essington, S. R. Palumbi, P. J. Mumby, and M. L. Pinsky. 2021. "Evolution and connectivity influence the persistence and recovery of coral reefs under climate change in the Caribbean, Southwest Pacific, and Coral Triangle." *Global Change Biology* 27(18):4307-4321. https://doi.org/10.1111/gcb.15725.

McParland, E. L., F. Wittmers, L. M. Bolaños, C. A. Carlson, R. Curry, S. J. Giovannoni, M. Michelsen, R. J. Parsons, M. C. Kido Soule, G. J. Swarr, B. Temperton, K. Vergin, A. Z. Worden, K. Longnecker, and

E. B. Kujawinski. 2024. "Seasonal exometabolites are regulated by essential microbial metabolisms in the oligotrophic ocean." *bioRxiv* 2024.03.05.583599. https://doi.org/10.1101/2024.03.05.583599.

Mecking, J. V., and S. S. Drijfhout. 2023. "The decrease in ocean heat transport in response to global warming." *Nature Climate Change* 13(11):1229. https://doi.org/10.1038/s41558-023-01829-8.

Melgar, D., and G. P. Hayes. 2019. "Characterizing large earthquakes before rupture is complete." *Science Advances* 5(5):eaav2032. https://doi.org/doi:10.1126/sciadv.aav2032.

Miles, T. N., D. Zhang, G. R. Foltz, J. A. Zhang, C. Meinig, F. Bringas, J. Triñanes, M. Le Hénaff, M. F. A. Vargas, S. Coakley, C. R. Edwards, D. Gong, R. E. Todd, M. J. Oliver, W. D. Wilson, K. Whilden, B. Kirkpatrick, P. Chardon-Maldonado, J. M. Morell, D. Hernandez, G. Kuska, C. D. Stienbarger, K. Bailey, C. Zhang, S. M. Glenn, and G. J. Goni. 2021. "Uncrewed Ocean gliders and saildrones support hurricane forecasting and research." *Oceanography* 34(4):78-81. https://www.jstor.org/stable/27217348.

Moftakhari, H. R., G. Salvadori, A. AghaKouchak, B. F. Sanders, and R. A. Matthew. 2017. "Compounding effects of sea level rise and fluvial flooding." *Proceedings of the National Academy of Sciences* 114(37):9785-9790. https://doi.org/doi:10.1073/pnas.1620325114.

Mooney, C., B. Dennis, K. Croww, and J. Muyskens. 2024. "The Drowning South—Where seas are rising at alarming speed." *The Washington Post*. https://www.washingtonpost.com/climate-environment/inter active/2024/southern-us-sea-level-rise-risk-cities (accessed March 20, 2025).

Moran, S. B. 2021. "Workforce development and leadership training for the new blue economy." In *Preparing a Workforce for the New Blue Economy*, edited by Liesl Hotaling and Richard W. Spinrad, pp. 407-416. North Holland: Elsevier.

Moran, S. B., M. M. Higgins, and D. E. Rosen. 2009. "Educating future business leaders in the strategic management of global change opportunities." *Management Education for Global Sustainability*, p. 227. Charlotte, NC: Information Age Publishing.

Mwampamba, T. H., B. N. Egoh, I. Borokini, and K. Njabo. 2022. "Challenges encountered when doing research back home: Perspectives from African conservation scientists in the diaspora." *Conservation Science and Practice* 4(5):e564. https://doi.org/10.1111/csp2.564.

NAML (National Association of Marine Laboratories). 2023. *Summary Report of the National Association of Marine Laboratories 2021 Member Survey February 2023.* https://www.naml.org/resources/reports/2023 %20Summary%20Report%20on%20the%202021%20NAML%20Survey.pdf (accessed March 20, 2025).

Narchi, N. E., G. G. M. Moura, and G. Leddy. 2024. "Blue economies and ocean grabbing in Latin America." *Latin American Perspectives* 51(3):5-25. https://doi.org/10.1177/0094582x241283774.

NASEM (National Academies of Science, Engineering, and Medicine). 2017a. *Sustaining Ocean Observations to Understand Future Changes in Earth's Climate*. Washington, DC: The National Academies Press.

NASEM. 2017b. *Acquisition and Operation of Polar Icebreakers: Fulfilling the Nation's Needs*. Washington, DC: The National Academies Press.

NASEM. 2018. *Sexual Harassment of Women: Climate, Culture, and Consequences in Academic Sciences, Engineering, and Medicine.* Edited by Paula A. Johnson, Sheila E. Widnall, and Frazier F. Benya. Washington, DC: The National Academies Press.

NASEM. 2020. *Earth System Predictability Research and Development: Proceedings of a Workshop–in Brief.* Edited by Kelly Oskvig and Amanda Staudt. Washington, DC: The National Academies Press.

NASEM. 2022a. *A Research Strategy for Ocean-based Carbon Dioxide Removal and Sequestration.* Washington, DC: The National Academies Press.

NASEM. 2022b. *Cross-Cutting Themes for U.S. Contributions to the UN Ocean Decade.* Washington, DC: The National Academies Press.

NASEM. 2022c. *Oil in the Sea IV: Inputs, Fates, and Effects.* Washington, DC: The National Academies Press.

NASEM. 2023. *Advancing Antiracism, Diversity, Equity, and Inclusion in STEMM Organizations: Beyond Broadening Participation.* Washington, DC: The National Academies Press.

NASEM. 2024a. "The State of the Science Address delivered by NAS President Marcia McNutt." https://www.nationalacademies.org/event/41687_06-2024_the-state-of-the-science (accessed March 20, 2025).

NASEM. 2024b. *Progress and Priorities in Ocean Drilling: In Search of Earth's Past and Future.* Washington, DC: The National Academies Press.

NASEM. 2024c. *Future Directions for Southern Ocean and Antarctic Nearshore and Coastal Research.* Washington, DC: The National Academies Press.

NASEM. 2024d. *Ocean Acoustics Education and Expertise.* Washington, DC: The National Academies Press.

Newman, A., N. Barlow, B. Brooks, D. Charlevoix, D. Foster, S. Schmidt, S. Webb, and M. Zumberge. 2022. *Workshop on the Alaska and Cascadia Seafloor Geodetic Community Experiment.* May 2-6, 2022, virtual. https://drive.google.com/file/d/1OI9kdZeIepyPUAQA130LDgvr6KYwc9bK/view (accessed March 20, 2025).

Newman, A., N. Bartlow, D. Schmidt, D. Charlevoix, J. Foster, B. Haines, E. Araki, R. Bürgmann, W. Chadwick, J. Gomberg, D. Melgar, L. Wallace, M. Wei, and W. Wilcock. 2021. *Future Directions in Seafloor Geodesy 2021 Community Workshop,* April 6-9, virtual. https://drive.google.com/file/d/1Smcd5a0EqO lSOSZeVHSqJNQij0XpF9Gx/view (accessed March 20, 2025).

NOAA (National Oceanic and Atmospheric Administration). 2022. *NOAA National Estuarine Research Reserve System Science Collaborative. Guide to Collaborative Science.* https://nerrssciencecollaborative.org/guide (accessed March 20, 2025).

NOAA. 2023. *The United States Ocean Acidification Action Plan.* https://oceanacidification.noaa.gov/wp-content/uploads/2024/01/Ocean-Acidification-Action-Plan.pdf (accessed March 20, 2025).

NOAA. 2025a. *Economics and Demographics.* Office for Coastal Management. https://coast.noaa.gov/states/fast-facts/economics-and-demographics.html (accessed August 14, 2024).

NOAA. 2025b. *Hurricane Costs.* Office for Coastal Management. https://coast.noaa.gov/states/fast-facts/hurricane-costs.html (accessed September 30, 2024).

Nooner, S. L., and W. W. Chadwick, Jr. 2016. "Inflation-predictable behavior and co-eruption deformation at Axial Seamount." *Science* 354(6318):1399-1403. https://doi.org/10.1126/science.aah4666.

Normile, D. 2024. "China's 'dreamy' new ship aims for Earth's mantle—and assumes ocean-drilling leadership." *Science*, December 11. https://www.science.org/content/article/china-s-dreamy-new-ship-aims-earth-s-mantle-and-assumes-ocean-drilling-leadership (accessed March 20, 2025).

Nowicki, M., T. DeVries, and D. A. Siegel. 2022. "Quantifying the carbon export and sequestration pathways of the ocean's biological carbon pump." *Global Biogeochemical Cycles* 36(3):e2021GB007083. https://doi.org/10.1029/2021GB007083.

NRC (National Research Council). 2009. *Science at Sea: Meeting Future Oceanographic Goals with a Robust Academic Research Fleet.* Washington, DC: The National Academies Press.

NRC. 2015. *Sea Change.* Washington, DC: The National Academies Press.

NSB (National Science Board). 2024. *A Changed Science and Engineering Landscape.* https://www.nsf.gov/nsb/publications/2024/changedlandscape.pdf?utm_medium=email&utm_source=govdelivery (accessed March 20, 2025).

NSF (National Science Foundation). n.d.-a. *About RISE.* https://new.nsf.gov/geo/rise/about (accessed September 16, 2024).

NSF. n.d.-b. *Budget, Performance and Financial Reporting.* https://new.nsf.gov/about/budget#financial-reporting-cd6 (March 20, 2025).

NSF. 2019. *MRI: Development of an Autonomous Four Dimensional Planktonic Microbial Environmental Sampling and Survey System: MiVEGAS.* https://www.nsf.gov/awardsearch/showAward?AWD_ID=1337601 (accessed December 18, 2024).

NSF. 2024a. *Expanding Geographic and Institutional Diversity in Computer and Information Science and engineering (CISE).* NSF 24-056. https://new.nsf.gov/funding/opportunities/dcl-expanding-geographic-institutional-diversity-computer (accessed March 20, 2025).

NSF. 2024b. *Expanding Geographic and Institutional Diversity in Social, Behavioral, and Economic Sciences (SBE).* NSF 24-079. https://new.nsf.gov/funding/opportunities/dcl-expanding-geographic-institutional-diversity-social (accessed March 20, 2025).

NSF. 2024c. *Formation of a New Subcommittee for a New Scientific Ocean Drilling Platform.* https://new.nsf.gov/geo/oce/updates/formation-subcommittee-new-scientific-ocean-drilling (accessed March 20, 2025).

NSF. 2024d. *Technology, Innovation and Partnerships: A Directorate at the U.S. National Science Foundation.* https://www.nsf.gov/awardsearch/showAward?AWD_ID=1946578 (accessed March 19, 2024).

OAP (Ocean Acidification Program). 2023. *Announcing $24.3M investment Advancing Marine Carbon Dioxide Removal Research*. National Oceanic and Atmospheric Administration. https://oceanacidification.noaa.gov/fy23-nopp-mcdr-awards (accessed August 8, 2024).

Obara, K., and A. Kato. 2016. "Connecting slow earthquakes to huge earthquakes." *Science* 353(6296):253-257. https://doi.org/10.1126/science.aaf1512.

Ober, K., and K. Waters. 2023. *Pacific Island Nations Seek Climate Solutions Outside of COP28*. https://www.usip.org/publications/2023/11/pacific-island-nations-seek-climate-solutions-outside-op28#:~:text=While%20the%20Pacific%20Islands%20are,more%20than%20double%20by%202100 (accessed August 8, 2024).

Ocean Research Advisory Panel. (2024). *Toward a National Ocean Data Strategy*. https://www.noaa.gov/sites/default/files/2024-09/508-compliant-Letter-and-ORAP-Report-001-2024.pdf (accessed March 20, 2025).

"Ocean temperatures surge, threatening worst coral bleaching event in history, scientists say." 2024. Fox News, May 17. https://www.foxnews.com/science/ocean-temperatures-surge-threatening-worst-coral-bleaching-event-history-scientists-say (accessed March 20, 2025).

OECD (Organisation for Economic Co-operation and Development). 2016. *The Ocean Economy in 2030*.Okazaki, R. R., A. J. Sutton, R. A. Feely, A. G. Dickson, S. R. Alin, C. L. Sabine, P. M. E. Bunje, and J. I. Virmani. 2017. "Evaluation of marine pH sensors under controlled and natural conditions for the Wendy Schmidt Ocean Health XPRIZE." *Limnology and Oceanography-Methods* 15(6):586-600. https://doi.org/10.1002/lom3.10189.

O'Leary, J. K., F. Micheli, L. Airoldi, C. Boch, G. De Leo, R. Elahi, F. Ferretti, N. A. J. Graham, S. Y. Litvin, N. H. Low, S. Lummis, K. J. Nickols, and J. Wong. 2017. "The resilience of marine ecosystems to climatic disturbances." *Bioscience* 67(3):208-220. https://doi.org/10.1093/biosci/biw161.

Omand, M. M., E. A. D'Asaro, C. M. Lee, M. J. Perry, N. Briggs, I. Cetinić, and A. Mahadevan. 2015. "Eddy-driven subduction exports particulate organic carbon from the spring bloom." *Science* 348(6231):222-225. https://doi.org/10.1126/science.1260062.

OOI (Ocean Observatories Initiative). 2025. Provided by Amber Coogon, Woods Hole Oceanographic Institution, Metrics for OOI 2.5, January 31. Available from the project public access file at https://www8.nationalacademies.org/pa/managerequest.aspx?key=DELS-OSB-22-04.

OPC (Ocean Policy Committee). 2023a. *Ocean Climate Action Plan*. Washington, DC: The White House. https://bidenwhitehouse.archives.gov/wp-content/uploads/2023/03/Ocean-Climate-Action-Plan_Final.pdf (accessed March 20, 2025).

OPC. 2023b. *Ocean Justice Strategy*. Washington, DC: The White House. https://bidenwhitehouse.archives.gov/wp-content/uploads/2023/12/Ocean-Justice-Strategy.pdf?cb=1701982354 (accessed March 20, 2025).

OPC. 2024. *National Strategy for a Sustainable Ocean Economy*. Washington, DC: The White House.

Orcutt, B. N., J. A. Bradley, W. J. Brazelton, E. R. Estes, J. M. Goordial, J. A. Huber, R. M. Jones, N. Mahmoudi, J. J. Marlow, S. Murdock, and M. Pachiadaki. 2020. "Impacts of deep-sea mining on microbial ecosystem services." *Limnology and Oceanography* 65(7):1489-1510. https://doi.org/10.1002/lno.11403.

Ortner, P. B. 2013. "Marine infrastructure: NSF fleet vital for ocean science." *Nature* 504(7479):218. https://doi.org/10.1038/504218a.

Oschlies, A. 2021. "A committed fourfold increase in ocean oxygen loss." *Nature Communications* 12(1):2307. https://doi.org/10.1038/s41467-021-22584-4.

Paine, R. T. 1966. "Food Web Complexity and Species Diversity." *The American Naturalist* 100(910):65-75. https://doi.org/10.1086/282400.

Palumbi, S. R., and B. D. Kessing. 1991. "Population biology of the Trans-Arctic Exchange: MTDNA sequence similarity between Pacific and Atlantic sea urchins." *Evolution* 45(8):1790-1805. https://doi.org/10.1111/j.1558-5646.1991.tb02688.x.

Palumbi, S. R., P. A. Sandifer, J. D. Allan, M. W. Beck, D. G. Fautin, M. J. Fogarty, B. S. Halpern, L. S. Incze, J.-A. Leong, E. Norse, J. J. Stachowicz, and D. H. Wall. 2009. "Managing for ocean biodiversity to sustain marine ecosystem services." *Frontiers in Ecology and the Environment* 7(4):204-211. https://doi.org/10.1890/070135.

Panet, I., S. Bonvalot, C. Narteau, D. Remy, and J.-M. Lemoine. 2018. "Migrating pattern of deformation prior to the Tohoku-Oki earthquake revealed by GRACE data." *Nature Geoscience* 11(5):367-373.

Pargett, D. M., J. M. Birch, C. M. Preston, J. P. Ryan, Y. Zhang, and C. A. Scholin. 2015. "Development of a mobile ecogenomic sensor." OCEANS 2015—MTS/IEEE Washington, October 19-22, 2015.

Paulus, E. 2021. "Shedding light on deep-sea biodiversity-a highly vulnerable habitat in the face of anthropogenic change." *Frontiers in Marine Science* 8. https://doi.org/10.3389/fmars.2021.667048.

Payne, A. E., M.-E. Demory, L. R. Leung, A. M. Ramos, C. A. Shields, J. J. Rutz, N. Siler, G. Villarini, A. Hall, and F. M. Ralph. 2020. "Responses and impacts of atmospheric rivers to climate change." *Nature Reviews Earth & Environment* 1(3):143-157. https://doi.org/10.1038/s43017-020-0030-5.

Payne, A. J., S. Nowicki, A. Abe-Ouchi, C. Agosta, P. Alexander, T. Albrecht, X. Asay-Davis, A. Aschwanden, A. Barthel, T. J. Bracegirdle, R. Calov, C. Chambers, Y. Choi, R. Cullather, J. Cuzzone, C. Dumas, T. L. Edwards, D. Felikson, X. Fettweis, B. K. Galton-Fenzi, H. Goelzer, R. Gladstone, N. R. Golledge, J. M. Gregory, R. Greve, T. Hattermann, M. J. Hoffman, A. Humbert, P. Huybrechts, N. C. Jourdain, T. Kleiner, P. K. Munneke, E. Larour, S. Le clec'h, V. Lee, G. Leguy, W. H. Lipscomb, C. M. Little, D. P. Lowry, M. Morlighem, I. Nias, F. Pattyn, T. Pelle, S. F. Price, A. Quiquet, R. Reese, M. Rückamp, N.-J. Schlegel, H. Seroussi, A. Shepherd, E. Simon, D. Slater, R. S. Smith, F. Straneo, S. Sun, L. Tarasov, L. D. Trusel, J. Van Breedam, R. van de Wal, M. van den Broeke, R. Winkelmann, C. Zhao, T. Zhang, and T. Zwinger. 2021. "Future Sea Level Change Under Coupled Model Intercomparison Project Phase 5 and Phase 6 Scenarios From the Greenland and Antarctic Ice Sheets." *Geophysical Research Letters* 48(16):e2020GL091741. https://doi.org/10.1029/2020GL091741.

Piecuch, C. G., and B. D. Hamlington. 2023. "Yesterday's high tide is today's new normal." *Earth's Future* 11(8):e2023EF003774. https://doi.org/10.1029/2023EF003774.

Pisani, J. 2024. "Storms again strike Texas, leaving hundreds of thousands without power." *The Wall Street Journal*, May 29. https://www.wsj.com/us-news/climate-environment/storms-again-strike-texas-leaving-hundreds-of-thousands-without-power-22fda307?mod=climate-environment_more_article_pos44 (accessed March 20, 2025).

Pohl, C. 2011. "What is progress in transdisciplinary research?" *Futures* 43(6):618-626. https://doi.org/10.1016/j.futures.2011.03.001.

Price, I., A. Sanchez-Gonzalez, F. Alet, T. R. Andersson, A. El-Kadi, D. Masters, T. Ewalds, J. Stott, S. Mohamed, P. Battaglia, R. Lam, and M. Willson. 2025. "Probabilistic weather forecasting with machine learning." *Nature* 637(8044):84-90. https://doi.org/10.1038/s41586-024-08252-9.

Pritchard, M. E., R. M. Allen, T. W. Becker, M. D. Behn, E. E. Brodsky, R. Bürgmann, C. Ebinger, J. T. Freymueller, M. Gerstenberger, and B. Haines. 2020. "New opportunities to study earthquake precursors." *Seismological Research Letters* 91(5):2444-2447.

PAME (Protection of the Arctic Marine Environment). 2019. "Meaningful engagement of Indigenous peoples and local communities in marine activities." *Report Findings for Policy Makers, Part II.* https://old.pame.is/index.php/document-library/pame-reports-new/pame-ministerial-deliverables/2019-11th-arctic-council-ministerial-meeting-rovaniemi-finland/425-meaningful-engagement-of-indigenous-peoples-and-local-communities-in-marine-activities-mema-part-ii-findings-for-policy-makers/file (accessed March 20, 2025).

Pulver, D. V. 2024. "Unprecedented ocean temperatures make this hurricane season especially dangerous." *USA Today*, June 2. https://www.usatoday.com/story/news/nation/2024/06/02/hot-atlantic-ocean-temperatures-fuel-hurricane-season-2024-danger/73889241007 (accessed March 20, 2025).

Rae, J. W. B., Y. G. Zhang, X. Liu, G. L. Foster, H. M. Stoll, and R. D. M. Whiteford. 2021. "Atmospheric CO_2 over the Past 66 Million Years from Marine Archives." *Annual Review of Earth and Planetary Sciences* 49:609-641. https://doi.org/10.1146/annurev-earth-082420-063026.

Rahmstorf, S. 2024. "Is the Atlantic overturning circulation approaching a tipping point?" *Oceanography* 37(3):16-29. https://www.jstor.org/stable/27333920.

Ralph, F. M., and M. D. Dettinger. 2011. "Storms, floods, and the science of atmospheric rivers." *EOS, Transactions American Geophysical Union* 92(32):265-266. https://doi.org/10.1029/2011EO320001.

Ranghieri, F., and M. Ishiwatari. 2014. *Learning From Megadisasters: Lessons from the Great East Japan Earthquake.* Washington, DC: World Bank Publications.

Rayadin, Y., and Z. Buřivalová. 2022. "What does it take to have a mutually beneficial research collaboration across countries?" *Conservation Science and Practice* 4(5):e528. https://doi.org/10.1111/csp2.528.

Reid, A. J., L. E. Eckert, J.-F. Lane, N. Young, S. G. Hinch, C. T. Darimont, S. J. Cooke, N. C. Ban, and A. Marshall. 2021. "'Two-eyed seeing': An Indigenous framework to transform fisheries research and management." *Fish and Fisheries* 22(2):243-261. https://doi.org/10.1111/faf.12516.

Renn, O. 2021. "Transdisciplinarity: Synthesis towards a modular approach." *Futures* 130:102744. https://doi.org/10.1016/j.futures.2021.102744.

Rex, M. A., and R. J. Etter. 2010. *Deep-sea biodiversity: pattern and scale*. Cambridge: Harvard University Press.

Rhoades, A., A. Jones, A. Srivastava, H. Huang, T. O'Brien, C. Patricola, P. Ullrich, M. Wehner, and Y. Zhou. 2020. "The shifting scales of western US landfalling atmospheric rivers under climate change." *Geophysical Research Letters* 47. https://doi.org/10.1029/2020GL089096.

Rippeth, T. P., and E. C. Fine. 2022. "Turbulent mixing in a changing Arctic Ocean." *Oceanography* 35(3/4):66-75. https://www.jstor.org/stable/27182697.

Ritchie, P. D. L., J. J. Clarke, P. M. Cox, and C. Huntingford. 2021. "Overshooting tipping point thresholds in a changing climate." *Nature* 592(7855):517-523. https://doi.org/10.1038/s41586-021-03263-2.

Robel, A. A., E. Wilson, and H. Seroussi. 2022. "Layered seawater intrusion and melt under grounded ice." *The Cryosphere* 16(2):451-469. https://doi.org/10.5194/tc-16-451-2022.

Roemmich, D., M. H. Alford, H. Claustre, K. Johnson, B. King, J. Moum, P. Oke, W. B. Owens, S. Pouliquen, S. Purkey, M. Scanderbeg, T. Suga, S. Wijffels, N. Zilberman, D. Bakker, M. Baringer, M. Belbeoch, H. C. Bittig, E. Boss, P. Calil, F. Carse, T. Carval, F. Chai, D. O. Conchubhair, F. d'Ortenzio, G. Dall'Olmo, D. Desbruyeres, K. Fennel, I. Fer, R. Ferrari, G. Forget, H. Freeland, T. Fujiki, M. Gehlen, B. Greenan, R. Hallberg, T. Hibiya, S. Hosoda, S. Jayne, M. Jochum, G. C. Johnson, K. Kang, N. Kolodziejczyk, A. Körtzinger, P. Y. Le Traon, Y. D. Lenn, G. Maze, K. A. Mork, T. Morris, T. Nagai, J. Nash, A. N. Garabato, A. Olsen, R. R. Pattabhi, S. Prakash, S. Riser, C. Schmechtig, C. Schmid, E. Shroyer, A. Sterl, P. Sutton, L. Talley, T. Tanhua, V. Thierry, S. Thomalla, J. Toole, A. Troisi, T. W. Trull, J. Turton, P. J. Velez-Belchi, W. Walczowski, H. L. Wang, R. Wanninkhof, A. F. Waterhouse, S. Waterman, A. Watson, C. Wilson, A. P. S. Wong, J. P. Xu, and I. Yasuda. 2019. *On the Future of Argo: A Global, Full-Depth, Multi-Disciplinary Array. Frontiers in Marine Science* 6. https://doi.org/10.3389/fmars.2019.00439.

Roquet, F., and C. Wunsch. 2022. "The Atlantic meridional overturning circulation and its hypothetical collapse." *Tellus A* 74:393-398.

Rossby, T., R. Curry, and J. Palter. 2017. "Packing Science into a Shipping Vessel". *Eos*, April 28. https://eos.org/science-updates/packing-science-into-a-shipping-vessel (accessed March 20, 2025).

Rossby, T., G. Siedler, and W. Zenk. 1995. "The Volunteer Observing Ship and Future Ocean Monitoring." *Bulletin of the American Meteorological Society* 76(1):5-12. https://doi.org/10.1175/1520-0477(1995)076<0005:TVOSAF>2.0.CO;2.

Rudnick, D. L., K. D. Zaba, R. E. Todd, and R. E. Davis. 2017. "A climatology of the California Current System from a network of underwater gliders." *Progress in Oceanography* 154:64-106. https://doi.org/10.1016/j.pocean.2017.03.002.

Ruiz, S., M. Metois, A. Fuenzalida, J. Ruiz, F. Leyton, R. Grandin, C. Vigny, R. Madariaga, and J. Campos. 2014. "Intense foreshocks and a slow slip event preceded the 2014 Iquique M_w 8.1 earthquake." *Science* 345(6201):1165-1169. https://doi.org/doi:10.1126/science.1256074.

Rusch, D. B., A. L. Halpern, G. Sutton, K. B. Heidelberg, S. Williamson, S. Yooseph, D. Wu, J. A. Eisen, J. M. Hoffman, K. Remington, K. Beeson, B. Tran, H. Smith, H. Baden-Tillson, C. Stewart, J. Thorpe, J. Freeman, C. Andrews-Pfannkoch, J. E. Venter, K. Li, S. Kravitz, J. F. Heidelberg, T. Utterback, Y.-H. Rogers, L. I. Falcón, V. Souza, G. Bonilla-Rosso, L. E. Eguiarte, D. M. Karl, S. Sathyendranath, T. Platt, E. Bermingham, V. Gallardo, G. Tamayo-Castillo, M. R. Ferrari, R. L. Strausberg, K. Nealson, R. Friedman, M. Frazier, and J. C. Venter. 2007. "The Sorcerer II Global Ocean Sampling Expedition: Northwest Atlantic through Eastern Tropical Pacific." *PLOS Biology* 5(3):e77. https://doi.org/10.1371/journal.pbio.0050077.

Russell, A. W., F. Wickson, and A. L. Carew. 2008. "Transdisciplinarity: Context, contradictions and capacity." *Futures* 40(5):460-472. https://doi.org/10.1016/j.futures.2007.10.005.

Sahneh, F., M. A. Balk, M. Kisley, C. K. Chan, M. Fox, B. Nord, E. Lyons, T. Swetnam, D. Huppenkothen, W. Sutherland, R. L. Walls, D. P. Quinn, T. Tarin, D. LeBauer, D. Ribes, D. P. Birnie, 3rd, C. Lushbough, E. Carr, G. Nearing, J. Fischer, K. Tyle, L. Carrasco, M. Lang, P. W. Rose, R. R. Rushforth, S. Roy, T. Matheson, T. Lee, C. T. Brown, T. K. Teal, M. Papes, S. Kobourov, and N. Merchant. 2021. "Ten simple rules to cultivate transdisciplinary collaboration in data science." *PLOS Computational Biology* 17(5):e1008879.

Sallée, J. B., E. P. Abrahamsen, C. Allaigre, M. Auger, H. Ayres, R. Badhe, J. Boutin, J. A. Brearley, C. de Lavergne, A. M. M. ten Doeschate, E. S. Droste, M. D. du Plessis, D. Ferreira, I. S. Giddy, B. Gülk, N. Gruber, M. Hague, M. Hoppema, S. A. Josey, T. Kanzow, M. Kimmritz, M. R. Lindeman, P. J. Llanillo, N. S. Lucas, G. Madec, D. P. Marshall, A. J. S. Meijers, M. P. Meredith, M. Mohrmann, P. M. S. Monteiro, C. Mosneron Dupin, K. Naeck, A. Narayanan, A. C. Naveira Garabato, S.-A. Nicholson, A. Novellino, M. Ödalen, S. Østerhus, W. Park, R. D. Patmore, E. Piedagnel, F. Roquet, H. S. Rosenthal, T. Roy, R. Saurabh, Y. Silvy, T. Spira, N. Steiger, A. F. Styles, S. Swart, L. Vogt, B. Ward, and S. Zhou. 2023. "Southern Ocean carbon and heat impact on climate." *Philosophical Transactions of the Royal Society A: Mathematical, Physical and Engineering Sciences* 381(2249):20220056. https://doi.org/10.1098/rsta.2022.0056.

Savoca, M. S., M. Kumar, Z. Sylvester, M. F. Czapanskiy, B. Meyer, J. A. Goldbogen, and C. M. Brooks. 2024. "Whale recovery and the emerging human-wildlife conflict over Antarctic krill." *Nature Communications* 15(1):7708. https://doi.org/10.1038/s41467-024-51954-x.

Seafarers International Research Centre. 2003. *Women Seafarers: Global Employment Policies and Practices.* Geneva: International Labour Organization.

Selkoe, K. A., T. Blenckner, M. R. Caldwell, L. B. Crowder, A. L. Erickson, T. E. Essington, J. A. Estes, R. M. Fujita, B. S. Halpern, M. E. Hunsicker, C. V. Kappel, R. P. Kelly, J. N. Kittinger, P. S. Levin, J. M. Lynham, M. E. Mach, R. G. Martone, L. A. Mease, A. K. Salomon, J. F. Samhouri, C. Scarborough, A. C. Stier, C. White, and J. Zedler. 2015. "Principles for managing marine ecosystems prone to tipping points." *Ecosystem Health and Sustainability* 1(5):1-18. https://doi.org/doi:10.1890/EHS14-0024.1.

Sellner, K. G., G. J. Doucette, and G. J. Kirkpatrick. 2003. "Harmful algal blooms: Causes, impacts and detection." *Journal of Industrial Microbiology and Biotechnology* 30(7):383-406. https://doi.org/10.1007/s10295-003-0074-9.

Seo, H., L. W. O'Neill, M. A. Bourassa, A. Czaja, K. Drushka, J. B. Edson, B. Fox-Kemper, I. Frenger, S. T. Gille, B. P. Kirtman, S. Minobe, A. G. Pendergrass, L. Renault, M. J. Roberts, N. Schneider, R. J. Small, A. Stoffelen, and Q. Wang. 2023. "Ocean Mesoscale and frontal-scale ocean–atmosphere interactions and influence on large-scale climate: A review." *Journal of Climate* 36(7):1981-2013. https://doi.org/10.1175/JCLI-D-21-0982.1.

Shackleton, S., D. Baggenstos, J. A. Menking, M. N. Dyonisius, B. Bereiter, T. K. Bauska, R. H. Rhodes, E. J. Brook, V. V. Petrenko, J. R. McConnell, T. Kellerhals, M. Häberli, J. Schmitt, H. Fischer, and J. P. Severinghaus. 2020. "Global Ocean Heat Content in the Last Interglacial." *Nature Geoscience* 13(1):77-81. https://doi.org/10.1038/s41561-019-0498-0.

Shah Walter, S. R., U. Jaekel, H. Osterholz, A. T. Fisher, J. A. Huber, A. Pearson, T. Dittmar, and P. R. Girguis. 2018. "Microbial decomposition of marine dissolved organic matter in cool oceanic crust." *Nature Geoscience* 11(5):334-339. https://doi.org/10.1038/s41561-018-0109-5.

Shen, Z. C., and W. B. Wu. 2024. "Ocean bottom distributed acoustic sensing for oceanic seismicity detection and seismic ocean thermometry." *Journal of Geophysical Research-Solid Earth* 129(3):e2023JB027799. https://doi.org/ARTN e2023JB027799.

Sher, D., D. Segrè, and M. J. Follows. 2024. "Quantitative principles of microbial metabolism shared across scales." *Nature Microbiology* 9(8):1940-1953. https://doi.org/10.1038/s41564-024-01764-0.

Shin, C. P., and W. D. Allmon. 2023. "How we study cryptic species and their biological implications: A case study from marine shelled gastropods." *Ecology and Evolution* 13(9):e10360. https://doi.org/10.1002/ece3.10360.

Shum, P., B. T. Barney, J. K. O'Leary, and S. R. Palumbi. 2019. "Cobble community DNA as a tool to monitor patterns of biodiversity within kelp forest ecosystems." *Molecular Ecology Resources* 19(6):1470-1485. https://doi.org/10.1111/1755-0998.13067.

Siderius, M., and J. Gebbie. 2019. "Environmental information content of ocean ambient noise." *Journal of the Acoustical Society of America* 146(3):1824. https://doi.org/10.1121/1.5126520.

Singh, J., P. Kerr, E. Hamburger, and A. Civilizations. 2016. *Media and Information Literacy: Reinforcing Human Rights, Countering Radicalization and Extremism (The MILID yearbook, 2016).* Paris: UNESCO Publishing.

Smith, C. R., V. Tunnicliffe, A. Colaço, J. C. Drazen, S. Gollner, L. A. Levin, N. C. Mestre, A. Metaxas, T. N. Molodtsova, T. Morato, A. K. Sweetman, T. Washburn, and D. J. Amon. 2020. "Deep-sea misconceptions cause underestimation of seabed-mining impacts." *Trends in Ecology & Evolution* 35(10):853-857. https://doi.org/10.1016/j.tree.2020.07.002.

Smythe, T. C., R. Thompson, and C. Garcia-Quijano. 2014. "The inner workings of collaboration in marine ecosystem-based management: A social network analysis approach." *Marine Policy* 50:117-125. https://doi.org/10.1016/j.marpol.2014.05.002.

Socquet, A., J. P. Valdes, J. Jara, F. Cotton, A. Walpersdorf, N. Cotte, S. Specht, F. Ortega-Culaciati, D. Carrizo, and E. Norabuena. 2017. "An 8-month slow slip event triggers progressive nucleation of the 2014 Chile megathrust." *Geophysical Research Letters* 44(9):4046-4053. https://doi.org/10.1002/2017GL073023.

Sproson, A. D., Y. Yokoyama, Y. Miyairi, T. Aze, and R. L. Totten. 2022. "Holocene melting of the West Antarctic Ice Sheet driven by tropical Pacific warming." *Nature Communications* 13(1):2434. https://doi.org/10.1038/s41467-022-30076-2.

Stachowicz, J. J., J. F. Bruno, and J. E. Duffy. 2007. "Understanding the effects of marine biodiversity on communities and ecosystems." *Annual Review of Ecology, Evolution, and Systematics* 38(1):739-766. https://doi.org/10.1146/annurev.ecolsys.38.091206.095659.

Stat, M., M. J. Huggett, R. Bernasconi, J. D. DiBattista, T. E. Berry, S. J. Newman, E. S. Harvey, and M. Bunce. 2017. "Ecosystem biomonitoring with eDNA: Metabarcoding across the tree of life in a tropical marine environment." *Scientific Reports* 7(1):12240. https://doi.org/10.1038/s41598-017-12501-5.

Steelman, T., A. Bogdan, C. Mantyka-Pringle, L. Bradford, M. G. Reed, S. Baines, J. Fresque-Baxter, T. Jardine, S. Shantz, R. Abu, K. Staples, E. Andrews, L. Bharadwaj, G. Strickert, P. Jones, K. Lindenschmidt, G. Poelzer, and D. D. Network. 2021. "Evaluating transdisciplinary research practices: Insights from social network analysis." *Sustainability Science* 16(2):631-645. https://doi.org/10.1007/s11625-020-00901-y.

Stefanoudis, P. V., W. Y. Licuanan, T. H. Morrison, S. Talma, J. Veitayaki, and L. C. Woodall. 2021. "Turning the tide of parachute science." *Current Biology* 31(4):R184-R185. https://doi.org/10.1016/j.cub.2021.01.029.

Stukel, M. R., and H. W. Ducklow. 2017. "Stirring up the biological pump: Vertical mixing and carbon export in the Southern Ocean." *Global Biogeochemical Cycles* 31(9):1420-1434. https://doi.org/10.1002/2017GB005652.

Su, Z., J. Wang, P. Klein, A. F. Thompson, and D. Menemenlis. 2018. "Ocean submesoscales as a key component of the global heat budget." *Nature Communications* 9(1):775. https://doi.org/10.1038/s41467-018-02983-w.

SOST (Subcommittee on Ocean Science and Technology Committee on Environment). 2018. *Science and Technology for America's Oceans: A Decadal Vision.* The White House. https://trumpwhitehouse.archives.gov/wp-content/uploads/2018/11/Science-and-Technology-for-Americas-Oceans-A-Decadal-Vision.pdf.

SOST. 2022. *Opportunities and Actions for Ocean Science and Technology (2022-2028).* Washington, DC: The White House. https://www.energy.gov/sites/default/files/2024-03/03-2022-SOST-Opportunities-and-Actions-for-Ocean-Science-and-Technology-2022-2028.pdf (accessed March 20, 2025).

SOST. 2024. *The National Ocean Biodiversity Strategy.* Washington, DC: The White House. https://bidenwhitehouse.archives.gov/wp-content/uploads/2024/06/NSTC_National-Ocean-Biodiversity-Strategy.pdf (accessed March 20, 2025).

Sunagawa, S., S. G. Acinas, P. Bork, C. Bowler, S. G. Acinas, M. Babin, P. Bork, E. Boss, C. Bowler, G. Cochrane, C. de Vargas, M. Follows, G. Gorsky, N. Grimsley, L. Guidi, P. Hingamp, D. Iudicone, O. Jaillon, S. Kandels, L. Karp-Boss, E. Karsenti, M. Lescot, F. Not, H. Ogata, S. Pesant, N. Poulton, J. Raes, C. Sardet, M. Sieracki, S. Speich, L. Stemmann, M. B. Sullivan, S. Sunagawa, P. Wincker, D. Eveillard, G. Gorsky, L. Guidi, D. Iudicone, E. Karsenti, F. Lombard, H. Ogata, S. Pesant, M. B. Sullivan, P. Wincker, C. de Vargas, and C. Tara Oceans. 2020. "Tara Oceans: Towards global ocean ecosystems biology." *Nature Reviews Microbiology* 18(8):428-445. https://doi.org/10.1038/s41579-020-0364-5.

Suttle, C. A. 2007. "Marine viruses—Major players in the global ecosystem." *Nature Reviews Microbiology* 5(10):801-812. https://doi.org/10.1038/nrmicro1750.

Szuwalski, C. S., K. Aydin, E. J. Fedewa, B. Garber-Yonts, and M. A. Litzow. 2023. "The collapse of eastern Bering Sea snow crab." *Science* 382(6668):306-310. https://doi.org/10.1126/science.adf6035.

Takahashi, M., M. Saccò, J. H. Kestel, G. Nester, M. A. Campbell, M. van der Heyde, M. J. Heydenrych, D. J. Juszkiewicz, P. Nevill, K. L. Dawkins, C. Bessey, K. Fernandes, H. Miller, M. Power, M. Mousavi-Derazmahalleh, J. P. Newton, N. E. White, Z. T. Richards, and M. E. Allentoft. 2023. "Aquatic environmental DNA: A review of the macro-organismal biomonitoring revolution." *Science of the Total Environment* 873:162322. https://doi.org/10.1016/j.scitotenv.2023.162322.

Tan, Y. J., M. Tolstoy, F. Waldhauser, and W. S. Wilcock. 2016. "Dynamics of a seafloor-spreading episode at the East Pacific Rise." *Nature* 540(7632):261-265. https://doi.org/10.1038/nature20116.

Teramura, A., K. Koeda, A. Matsuo, M. P. Sato, H. Senou, H. C. Ho, Y. Suyama, K. Kikuchi, and S. Hirase. 2022. "Assessing the effectiveness of DNA barcoding for exploring hidden genetic diversity in deep-sea fishes." *Marine Ecology Progress Series* 701:83-98. https://www.int-res.com/abstracts/meps/v701/p83-98/.

Terhaar, J., T. L. Frolicher, and F. Joos. 2021. "Southern Ocean anthropogenic carbon sink constrained by sea surface salinity." *Science Advances* 7(18):eabd5964. https://doi.org/10.1126/sciadv.abd5964.

The Cenozoic CO_2 Proxy Integration Project (CenCO2PIP) Consortium. 2023. "Toward a Cenozoic history of atmospheric CO_2." *Science* 382(6675):eadi5177. https://doi.org/10.1126/science.adi5177.

The IMBIE team, S., Andrew, E. Ivins, E. Rignot, B. Smith, M. van den Broeke, I. Velicogna, P. Whitehouse, K. Briggs, I. Joughin, G. Krinner, S. Nowicki, T. Payne, T. Scambos, N. Schlegel, G. A, C. Agosta, A. Ahlstrøm, G. Babonis, V. Barletta, A. Blazquez, J. Bonin, B. Csatho, R. Cullather, D. Felikson, X. Fettweis, R. Forsberg, H. Gallee, A. Gardner, L. Gilbert, A. Groh, B. Gunter, E. Hanna, C. Harig, V. Helm, A. Horvath, M. Horwath, S. Khan, K. K. Kjeldsen, H. Konrad, P. Langen, B. Lecavalier, B. Loomis, S. Luthcke, M. McMillan, D. Melini, S. Mernild, Y. Mohajerani, P. Moore, J. Mouginot, G. Moyano, A. Muir, T. Nagler, G. Nield, J. Nilsson, B. Noel, I. Otosaka, M. E. Pattle, W. R. Peltier, N. Pie, R. Rietbroek, H. Rott, L. Sandberg-Sørensen, I. Sasgen, H. Save, B. Scheuchl, E. Schrama, L. Schröder, K.-W. Seo, S. Simonsen, T. Slater, G. Spada, T. Sutterley, M. Talpe, L. Tarasov, W. J. van de Berg, W. van der Wal, M. van Wessem, B. D. Vishwakarma, D. Wiese, and B. Wouters. 2018. "Mass balance of the Antarctic Ice Sheet from 1992 to 2017." *Nature* 558(7709):219-222.

The IMBIE team, S., Andrew, E. Ivins, E. Rignot, B. Smith, M. van den Broeke, I. Velicogna, P. Whitehouse, K. Briggs, I. Joughin, G. Krinner, S. Nowicki, T. Payne, T. Scambos, N. Schlegel, G. A, C. Agosta, A. Ahlstrøm, G. Babonis, V. R. Barletta, A. A. Bjørk, A. Blazquez, J. Bonin, W. Colgan, B. Csatho, R. Cullather, M. E. Engdahl, D. Felikson, X. Fettweis, R. Forsberg, A. E. Hogg, H. Gallee, A. Gardner, L. Gilbert, N. Gourmelen, A. Groh, B. Gunter, E. Hanna, C. Harig, V. Helm, A. Horvath, M. Horwath, S. Khan, K. K. Kjeldsen, H. Konrad, P. L. Langen, B. Lecavalier, B. Loomis, S. Luthcke, M. McMillan, D. Melini, S. Mernild, Y. Mohajerani, P. Moore, R. Mottram, J. Mouginot, G. Moyano, A. Muir, T. Nagler, G. Nield, J. Nilsson, B. Noël, I. Otosaka, M. E. Pattle, W. R. Peltier, N. Pie, R. Rietbroek, H. Rott, L. Sandberg Sørensen, I. Sasgen, H. Save, B. Scheuchl, E. Schrama, L. Schröder, K.-W. Seo, S. B. Simonsen, T. Slater, G. Spada, T. Sutterley, M. Talpe, L. Tarasov, W. J. van de Berg, W. van der Wal, M. van Wessem, B. D. Vishwakarma, D. Wiese, D. Wilton, T. Wagner, B. Wouters, and J. Wuite. 2020. "Mass balance of the Greenland Ice Sheet from 1992 to 2018." *Nature* 579(7798):233-239. https://doi.org/10.1038/s41586-019-1855-2.

Thornton, T. F., and A. M. Scheer. 2012. "Collaborative engagement of local and traditional knowledge and science in marine environments: A review." *Ecology and Society* 17(3). http://www.jstor.org/stable/26 269064.

Thurber, A. R., A. K. Sweetman, B. E. Narayanaswamy, D. O. B. Jones, J. Ingels, and R. L. Hansman. 2014. "Ecosystem function and services provided by the deep sea." *Biogeosciences* 11(14):3941-3963. https://doi.org/10.5194/bg-11-3941-2014.

Tolstoy, M., J. P. Cowen, E. T. Baker, D. J. Fornari, K. H. Rubin, T. M. Shank, F. Waldhauser, D. R. Bohnenstiehl, D. W. Forsyth, R. C. Holmes, B. Love, M. R. Perfit, R. T. Weekly, S. A. Soule, and B. Glazer. 2006. "A sea-floor spreading event captured by seismometers." *Science* 314(5807):1920-1922. https://doi.org/10.1126/science.1133950.

Toohey, G. 2024. "'Rivers in the sky' have drenched California, yet even more extreme rains are possible." *Los Angeles Times*, April 25. https://www.latimes.com/environment/story/2024-04-25/atmospheric-rivers-could-pound-california-with-more-extreme-rain (accessed March 20, 2025).

Trembath-Reichert, E., S. R. Shah Walter, M. A. F. Ortiz, P. D. Carter, P. R. Girguis, and J. A. Huber. 2021. "Multiple carbon incorporation strategies support microbial survival in cold subseafloor crustal fluids." *Science Advances* 7(18):eabg0153. https://doi.org/10.1126/sciadv.abg0153.

UN (United Nations). n.d. "Department of economic and social affairs: Indigenous peoples." https://www.un.org/development/desa/indigenouspeoples/climate-change.html (accessed Sept 27, 2024).

UN. 2019. *Climate Justice.* https://www.un.org/sustainabledevelopment/blog/2019/05/climate-justice (accessed Sept 17, 2024).

U.S. Census Bureau. 2019. "Coastline America." https://www.census.gov/content/dam/Census/library/visualiz ations/2019/demo/coastline-america.pdf (accessed March 20, 2025).

van Denderen, P. D., C. M. Petrik, C. A. Stock, K. H. Andersen, and A. Bates. 2021. "Emergent global biogeography of marine fish food webs." *Global Ecology and Biogeography* 30(9):1822-1834. https://doi.org/10.1111/geb.13348.

van Westen, R. M., M. Kliphuis, and H. A. Dijkstra. 2024. "Physics-based early warning signal shows that AMOC is on tipping course." *Science Advances* 10(6):eadk1189. https://doi.org/10.1126/sciadv.adk1189.

Varbanov, L. 2023. "China's rising tide: Expanding investment in blue finance." World Economic Forum. https://www.weforum.org/stories/2023/06/chinas-rising-tide-expanding-investment-in-blue-finance-amn c23 (accessed March 20, 2025).

Vasileiadou, K., E. Chatzinikolaou, S. Klayn, C. Pavloudi, and S. Reizopoulou. 2024. "Editorial: Marine biodiversity hotspots – challenges and resilience." *Frontiers in Marine Science* 11. https://doi.org/10.3389/fmars.2024.1338242.

von Schuckmann, K., A. Minière, F. Gues, F. J. Cuesta-Valero, G. Kirchengast, S. Adusumilli, F. Straneo, M. Ablain, R. P. Allan, P. M. Barker, H. Beltrami, A. Blazquez, T. Boyer, L. Cheng, J. Church, D. Desbruyeres, H. Dolman, C. M. Domingues, A. García-García, D. Giglio, J. E. Gilson, M. Gorfer, L. Haimberger, M. Z. Hakuba, S. Hendricks, S. Hosoda, G. C. Johnson, R. Killick, B. King, N. Kolodziejczyk, A. Korosov, G. Krinner, M. Kuusela, F. W. Landerer, M. Langer, T. Lavergne, I. Lawrence, Y. Li, J. Lyman, F. Marti, B. Marzeion, M. Mayer, A. H. MacDougall, T. McDougall, D. P. Monselesan, J. Nitzbon, I. Otosaka, J. Peng, S. Purkey, D. Roemmich, K. Sato, K. Sato, A. Savita, A. Schweiger, A. Shepherd, S. I. Seneviratne, L. Simons, D. A. Slater, T. Slater, A. K. Steiner, T. Suga, T. Szekely, W. Thiery, M.-L. Timmermans, I. Vanderkelen, S. E. Wjiffels, T. Wu, and M. Zemp. 2023. "Heat stored in the Earth system 1960–2020: Where does the energy go?" *Earth System Science Data* 15(4):1675-1709. https://doi.org/10.5194/essd-15-1675-2023.

Vonnahme, T. R., M. Molari, F. Janssen, F. Wenzhöfer, M. Haeckel, J. Titschack, and A. Boetius. 2020. "Effects of a deep-sea mining experiment on seafloor microbial communities and functions after 26 years." *Science Advances* 6(18):eaaz5922. https://doi.org/10.1126/sciadv.aaz5922.

Voosen, P. 2024. "Seafloor fiber-optic cables become sensor stations." *Science* 383(6688):1166-1167. https://doi.org/10.1126/science.adp1965.

Waite, K. A., P. J. Pronovost, and J. S. Barnholtz-Sloan. 2023. "Critical partnerships: How to develop a trans-disciplinary research team." *Cancers* 15(20):5078. https://doi.org/ARTN 507810.3390/cancers152050 78.

Walczak, M. H., A. C. Mix, E. A. Cowan, S. Fallon, L. K. Fifield, J. R. Alder, J. Du, B. Haley, T. Hobern, J. Padman, S. K. Praetorius, A. Schmittner, J. S. Stoner, and S. D. Zellers. 2020. "Phasing of millennial-scale climate variability in the Pacific and Atlantic Oceans." *Science* 370(6517):716-720. https://doi.org/10.1126/science.aba7096.

Walker, J. S., T. Li, T. A. Shaw, N. Cahill, D. C. Barber, M. J. Brain, R. E. Kopp, A. D. Switzer, and B. P. Horton. 2023. "A 5000-year record of relative sea-level change in New Jersey, USA." *The Holocene* 33(2):167-180. https://doi.org/10.1177/09596836221131696.

Wallis, B. J., A. E. Hogg, J. M. van Wessem, B. J. Davison, and M. R. van den Broeke. 2023. "Widespread seasonal speed-up of west Antarctic Peninsula glaciers from 2014 to 2021." *Nature Geoscience* 16(3):231-237. https://doi.org/10.1038/s41561-023-01131-4.

Wang, P., G. Hu, H. Wang, Y. Ge, Y. Ma, Z. Yu, X. Wang, and M. Fox. 2024. "Onset of Mid-Pleistocene glaciation in the Eastern Himalayan syntaxis." *Communications Earth & Environment* 5(1):346. https://doi.org/10.1038/s43247-024-01517-1.

Watson, J. E., O. Venter, J. Lee, K. R. Jones, J. G. Robinson, H. P. Possingham, and J. R. Allan. 2018. *Protect the Last of the Wild.* London: Nature Publishing Group.

Webster, S. E., E. C. Donovan, E. Chudoba, C. D. Miller Hesed, M. Paolisso, and W. C. Dennison. 2022. "Identifying and harmonizing the priorities of stakeholders in the Chesapeake Bay environmental monitoring community." *Current Research in Environmental Sustainability* 4:100155. https://doi.org/10.1016/j.crsust.2022.100155.

Whalen, C. B., C. de Lavergne, A. C. N. Garabato, J. M. Klymak, J. A. MacKinnon, and K. L. Sheen. 2020. "Internal wave-driven mixing: governing processes and consequences for climate." *Nature Reviews Earth & Environment* 1(11):606-621. https://doi.org/10.1038/s43017-020-0097-z.

White, C., K. A. Selkoe, J. Watson, D. A. Siegel, D. C. Zacherl, and R. J. Toonen. 2010. "Ocean currents help explain population genetic structure." *Proceedings of the Royal Society B: Biological Sciences* 277(1688):1685-1694. https://doi.org/10.1098/rspb.2009.2214.

Wilcock, W. S., M. Tolstoy, F. Waldhauser, C. Garcia, Y. J. Tan, D. R. Bohnenstiehl, J. Caplan-Auerbach, R. P. Dziak, A. F. Arnulf, and M. E. Mann. 2016. "Seismic constraints on caldera dynamics from the 2015 Axial Seamount eruption." *Science* 354(6318):1395-1399. https://doi.org/10.1126/science.aah5563.

Wilcock, W. S. D., S. Abadi, and B. P. Lipovsky. 2023. "Distributed acoustic sensing recordings of low-frequency whale calls and ship noise offshore Central Oregon." *JASA Express Letters* 3(2):026002. https://doi.org/10.1121/10.0017104.

Williamson, A. L., D. Melgar, B. W. Crowell, D. Arcas, T. I. Melbourne, Y. Wei, and K. Kwong. 2020. "Toward near-field tsunami forecasting along the Cascadia Subduction Zone using rapid GNSS source models." *Journal of Geophysical Research: Solid Earth* 125(8):e2020JB019636. https://doi.org/10.1029/2020JB019636.

Wilson, J. D., O. Andrews, A. Katavouta, F. de Melo Virissimo, R. M. Death, M. Adloff, C. A. Baker, B. Blackledge, F. W. Goldsworth, A. T. Kennedy-Asser, Q. Liu, K. R. Sieradzan, E. Vosper, and R. Ying. 2022. "The biological carbon pump in CMIP6 models: 21st century trends and uncertainties." *Proceedings of the National Academy of Sciences* 119(29):e2204369119. https://doi.org/10.1073/pnas.2204369119.

Wirtz, K., and S. L. Smith. 2020. "Vertical migration by bulk phytoplankton sustains biodiversity and nutrient input to the surface ocean." *Scientific Reports* 10(1):1142. https://doi.org/10.1038/s41598-020-57890-2.

Women in Ocean Science. 2021. *Sexual Harassment in Marine Science.* https://www.womeninoceanscience.com/sexual-harassment (accessed March 20, 2025).

Wong, T. E., A. M. R. Bakker, and K. Keller. 2017. "Impacts of Antarctic fast dynamics on sea-level projections and coastal flood defense." *Climatic Change* 144(2):347-364. https://doi.org/10.1007/s10584-017-2039-4.

Worcester, P., M. Dzieciuch, and H. Sagen. 2020. "Ocean acoustics in the rapidly changing Arctic." *Acoustics Today* 16(1). https://doi.org/10.1121/AT.2020.16.1.55.

Worm, B., E. B. Barbier, N. Beaumont, J. E. Duffy, C. Folke, B. S. Halpern, J. B. Jackson, H. K. Lotze, F. Micheli, S. R. Palumbi, E. Sala, K. A. Selkoe, J. J. Stachowicz, and R. Watson. 2006. "Impacts of biodiversity loss on ocean ecosystem services." *Science* 314(5800):787-90. https://doi.org/10.1126/science.1132294.

Wu, S., L. Lembke-Jene, F. Lamy, H. W. Arz, N. Nowaczyk, W. Xiao, X. Zhang, H. C. Hass, J. Titschack, X. Zheng, J. Liu, L. Dumm, B. Diekmann, D. Nurnberg, R. Tiedemann, and G. Kuhn. 2021. "Orbital- and millennial-scale Antarctic Circumpolar Current variability in Drake Passage over the past 140,000 years." *Nature Communications* 12(1):3948. https://doi.org/10.1038/s41467-021-24264-9.

Wu, W., and A. Mahadevan. 2024. "Air-sea turbulent heat flux affects oceanic lateral eddy heat transport." *Geophysical Research Letters* 51(21):e2024GL110459. https://doi.org/10.1029/2024GL110459.

Wu, W., Z. Zhan, S. Peng, S. Ni, and J. Callies. 2020. "Seismic ocean thermometry." *Science* 369(6510):1510-1515. https://doi.org/10.1126/science.abb9519.

Xia, C., C. Liu, Z. Cai, H. Wu, Q. Li, and M. Gao. 2024. "Tracking SO2 plumes from the Tonga volcano eruption with multi-satellite observations." *Iscience* 27(4).

Xu, G. Y., W. W. Chadwick, W. S. D. Wilcock, K. G. Bemis, and J. Delaney. 2018. "Observation and modeling of hydrothermal response to the 2015 eruption at Axial Seamount, northeast Pacific." *Geochemistry Geophysics Geosystems* 19(8):2780-2797. https://doi.org/10.1029/2018gc007607.

Yang, J., S. C. Riser, J. A. Nystuen, W. E. Asher, and A. T. Jessup. 2015. "Regional rainfall measurements using the passive aquatic listener during the SPURS field campaign." *Oceanography* 28(1):124-133. https://doi.org/10.5670/oceanog.2015.10.

Yasuhara, M., H. Doi, C.-L. Wei, R. Danovaro, and S. E. Myhre. 2016. "Biodiversity–ecosystem functioning relationships in long-term time series and palaeoecological records: Deep sea as a test bed." *Philosophical Transactions of the Royal Society B: Biological Sciences* 371(1694):20150282. https://doi.org/10.1098/rstb.2015.0282.

Yates, K. K., C. Turley, B. M. Hopkinson, A. E. Todgham, J. N. Cross, H. Greening, P. Williamson, R. Van Hooidonk, D. D. Deheyn, and Z. Johnson. 2015. "Transdisciplinary science a path to understanding the interactions among ocean acidification, ecosystem, and society." *Oceanography* 28(2):212-225. https://doi.org/10.5670/oceanog.2015.43.

Yokota, Y., T. Ishikawa, and S.-i. Watanabe. 2018. "Seafloor crustal deformation data along the subduction zones around Japan obtained by GNSS-A observations." *Scientific Data* 5(1):180182. https://doi.org/10.1038/sdata.2018.182.

Yu, L. 2019. "Global air–sea fluxes of heat, fresh water, and momentum: Energy budget closure and unanswered questions." *Annual Review of Marine Science* 11:227-248. https://doi.org/10.1146/annurev-marine-010816-060704.

Zerkel, E. 2024. "Watch the glacier outburst that sent a surge of water into Juneau, causing 'unprecedented' flooding." *CNN*, August 7. https://www.cnn.com/2024/08/07/climate/alaska-juneau-flooding-glacier-outburst-climate/index.html.

Zhang, D., A. M. Chiodi, C. Zhang, G. R. Foltz, M. F. Cronin, C. W. Mordy, J. Cross, E. D. Cokelet, J. A. Zhang, C. Meinig, N. Lawrence-Slavas, P. J. Stabeno, and R. Jenkins. 2023. "Observing Extreme Ocean and Weather Events Using Innovative Saildrone Uncrewed Surface Vehicles." *Oceanography* 36(2/3):70-77. https://www.jstor.org/stable/27257882 (accessed March 20, 2025).

Zhong, R., and M. Rojanasakul. 2024. "How close are the planet's climate tipping points?" *The New York Times* August 11. https://www.nytimes.com/interactive/2024/08/11/climate/earth-warming-climate-tipping-points.html?smid=nytcore-ios-share&referringSource=articleShare&sgrp=c-cb (accessed March 20, 2025).

Appendix A
Committee Biographies

H. Tuba Özkan-Haller (*Co-Chair*) is dean and professor at Oregon State University, College of Earth, Ocean, and Atmospheric Sciences. She has spent most of her career at the intersection of physical oceanography, marine geology, and coastal and ocean engineering. Özkan-Haller's research focuses on the use of numerical, field, laboratory, and analytical approaches to arrive at a predictive understanding of ocean waves, circulation, and coastal change. Özkan-Haller is the recipient of the Office of Naval Research Young Investigator Award, the Outstanding Faculty Member Award at the University of Michigan, and the Pattullo Award for Excellence in Teaching at Oregon State University. She currently serves on two federal advisory committees, namely the Board on Coastal Engineering Research and the Hydrographic Survey Review Panel. Özkan-Haller has been engaging extensively with NASEM, having served two two terms on NASEM's Ocean Studies Board, served as a member of the Marine and Hydrokinetic Energy Assessment Committee, and as chair of the Committee on Long-Term Coastal Zone Dynamics. In April 2021, she published the article "It's Time to Invest in Curiosity" in the *Fair Observer*. Özkan-Haller received a B.S. in civil engineering from Boğaziçi University, Turkey, and an M.C.E. and Ph.D. in civil engineering from the University of Delaware.

James (Jim) A. Yoder (*Co-Chair*) is dean emeritus at the Woods Hole Oceanographic Institution (WHOI) and professor emeritus at the University of Rhode Island (URI). His first academic position was as a researcher at the Skidaway Institute of Oceanography, participating in an interdisciplinary study of the southeastern U.S. continental shelf. Yoder joined the URI faculty in 1989, where he studied regional to global distributions of phytoplankton biomass and productivity using satellite and aircraft measurements. In 2005, he moved to WHOI, where he served as vice president for academic programs and dean. During his academic career, Yoder held temporary assignments as program officer at the headquarters of the National Aeronautics and Space Administration and as director of the National Science Foundation's Division of Ocean Sciences. He was selected fellow of The Oceanography Society in 2012 and fellow of the American Association for the Advancement of Science in 2019. He previously served on the Ocean Studies Board and as a member of the 2013–2015 Committee on the Decadal Survey of Ocean Sciences and as chair of the Committee on Catalyzing Opportunities for Research in the Earth Sciences (CORES): A Decadal Survey for NSF's Division of Earth Sciences. Yoder serves on the Corporation of the Woods Hole Oceanographic Institution. He received a B.A. from DePauw University and an M.S. and Ph.D. in oceanography from the University of Rhode Island.

Lihini Aluwihare is professor in marine chemistry and geochemistry at the University of California, San Diego, Scripps Institution of Oceanography. She is a chemical oceanographer who studies the cycling of carbon and nitrogen in the oceans using light isotope tools and organic matter chemical characterization. Aluwihare's work focuses around developing new analytical tools and research frameworks to read the messages encoded in molecules that maintain microbial life, facilitate ecosystem interactions, and contribute to long-term carbon and nutrient storage. She is also interested in the distribution and cycling of anthropogenic compounds in coastal environments. Aluwihare's career in academia has been guided by a need to build a community of scholars that adequately represents the interests and experiences of the broader population. She received a B.S. in chemistry and philosophy from Mount Holyoke College and a Ph.D. from the Massachusetts Institute of Technology–Woods Hole Oceanographic Institution Joint Program in Oceanography.

Mona Behl is associate director of Georgia Sea Grant at the University of Georgia, where she also holds public service and academic appointments. She is also a nonresidential policy fellow with the American

Meteorological Society (AMS). Behl's research interests include assisting coastal communities in managing the impacts of extreme weather and climate change, increase access and opportunity in Earth and environmental studies, and preparing people for the future of work. She co-chairs the Mentoring Physical Oceanography Women to Increase Retention program. She also directs an NSF-funded research coordination network that is focused on studying the impact of climate induces human mobility. Behl co-founded the AMS Early Carrer Leadership Academy and Sea Grant's Community Engaging Internship program. She is the vice-chair for UCAR's Community Program external advisory committee, is president-elect of the Earth Science Women's Network, and chair-elect of AMS's culture and inclusion cabinet. From 2022-2025, she served on the council of The Oceanography Society and the AMS. Behl is a recipient of Sea Grant's President Award and an Ocean Decade Champion. She received a B.S. and M.S in physics from Panjab University, India, and a Ph.D. in physical oceanography from Florida State University.

Mark D. Behn is a professor in the Morrissey College of Arts and Sciences at Boston College. His research investigates the dynamics of earth deformation in glacial, marine, and terrestrial environments through a wide range of geophysical techniques. These techniques include the development of geodynamic models that relate laboratory-based rheologic and petrologic models to the large-scale behavior of the earth, which are then applied to a spectrum of problems from basic science to societally relevant issues. Behn's research interests include dynamics of faulting, magmatism, and surface processes at midocean ridges and continental rifts; seismic anisotropy and imaging of sub-asthenospheric mantle flow; evolution of the continental crust; and ice sheet dynamics. He is an editor of the Journal of Geophysical Research—Solid Eath, co-chair for the SZ4D Modeling Collaboratory for Subduction, and former fellow of the Woods Hole Oceanographic Institution Deep Ocean Exploration Institute. Behn received a B.S. in geology from Bates College and a Ph.D. in marine geophysics from the Massachusetts Institute of Technology–Woods Hole Oceanographic Institution Joint Program.

Brad deYoung is executive director of the Pacific node of the Canadian Integrated Ocean Observing System (CIOOS) and professor emeritus and Robert A. Bartlett professor of oceanography at Memorial University. In addition to participating in national and international observing programs, such as the Overturning in the Subpolar North Atlantic Program, he has also been engaged in developing links to public policy and exploring opportunities to connect science, society, and economy. DeYoung served for a decade on the Canadian Fisheries Resource Conservation Council, advising the minister of fisheries and oceans on fisheries policy and management. He is working with ocean gliders to make year-round measurements in the Northwest Atlantic and is helping to develop the CIOOS observing network to support data access and the provision of new information services. DeYoung received a Ph.D. in physical oceanography from the University of British Columbia.

Carlos Garcia-Quijano holds a joint appointment as professor in the Department of Sociology and Anthropology and the Department of Marine Affairs at the University of Rhode Island. He has a special interest in how human cognition, culture, and society influence the interaction between people and the nonhuman environment, as well as who bears the impacts and responsibility for environmental problems. Garcia-Quijano's research involves a comparative study of cultural aspects of coastal use and dependence to reach a more comprehensive understanding of human well-being and adaptations as they relate to the use of coastal environments and resources, including local knowledge, resource management, and adaptations to species translocations. He received a B.S. in biology, an M.S. in geology and reef paleoecology from the University of Puerto Rico, and a Ph.D. in ecological and environmental anthropology from the University of Georgia.

Peter Girguis is professor of organismic and evolutionary biology and co-director of the Microbial Sciences Initiative at Harvard University. He is also adjunct professor in the Woods Hole Oceanographic Institution's Applied Ocean Physics and Engineering group. Girguis is a microbiologist, biogeochemist, and technologist who studies how animals and microbes in the deep sea influence biogeochemical cycles. He is also known for developing novel "open-design" instruments, such as underwater mass spectrometers and microbial samplers, which he strives to make available to the broadest research community with the goal of democratizing science around the world. Girguis was a National Science Foundation RIDGE 2000 distinguished lecturer, a Merck & Co. Innovative Research Awardee, a recipient of the 2007 and 2011 Lindbergh Foundation Award for Science & Sustainability, the 2018 Lowell Thomas Award for groundbreaking advances in marine science and technology, and the 2020 Petra Shattuck Award for Distinguished Teaching. He was recently named a Gordon and Betty Moore Foundation investigator. Girguis is a member of the National Oceanic and Atmospheric Administration's Ocean Exploration Advisory Board. He was previously a member of the National Academies of Sciences, Engineering, and Medicine's First Indian-American Frontiers of Science Symposium. Girguis received a B.S. from the University of California, Los Angeles, and a Ph.D. from the University of California, Santa Barbara, and completed postdoctoral research at the Monterey Bay Aquarium Research Institute.

Leila J. Hamdan serves as associate vice president for research, coastal operations and professor in the School of Ocean Science and Engineering at the University of Southern Mississippi. Her research centers on marine microbial biogeography and exploring natural and human-made features on the seabed that shape coastal to deep-sea ecosystems. Hamdan is the lead on the National Science Foundation award for the operation of the future Regional Class Research Vessel *Gilbert R. Mason* and was chief scientist on 25 oceanographic research expeditions. She is past president of the Coastal and Estuarine Research Federation. Hamdan received the National Oceanographic Partnership Program Excellence in Partnering Award in 2017 for leadership of a team of scientists studying impacts of the *Deepwater Horizon* oil spill. She received a B.S. in biology from Rowan University of New Jersey and an M.S. and Ph.D. from George Mason University and completed postdoctoral training at the Naval Research Laboratory.

Marcia J. Isakson is director of the Signal and Information Sciences Laboratory (SISL) at Applied Research Laboratories, the University of Texas at Austin (ARL:UT). SISL is involved in a broad spectrum of research including anti-submarine warfare, systems engineering, geospatial remote sensing, quantum information sciences, content understanding, nuclear surety, aeroacoustics, hypersonics, modeling, and simulation with a diverse sponsor base that includes the Department of Defense, the national intelligence community, National Aeronautics and Space Administration, Department of Energy, and the National Science Foundation. Isakson's research interests include the effects of the underwater environment on ocean acoustic propagation. She is a fellow and former president of the Acoustical Society of America (ASA) and taught graduate courses in underwater acoustics at the University of Texas at Austin from 2009–2018. She is currently a member of the Ocean Studies Board at the National Academies for Science, Engineering, and Medicine, a member of the U.S. Committee for the UN Decade of Ocean Science and a member of the Decadal Survey of Ocean Sciences for the National Science Foundation. Isakson received a B.S. in engineering physics and mathematics from the United States Military Academy at West Point in 1992 and was awarded a Hertz Foundation Fellowship upon graduation. She completed a master's degree and Ph.D. in physics from the University of Texas at Austin.

Jason Link is senior scientist for ecosystems with the National Oceanic and Atmospheric Administration's National Marine Fisheries Service, leading efforts to support development of ecosystem-based management plans and activities throughout the agency. He has held many past adjunct positions and currently holds an adjunct faculty position at the School for Marine Science and Technology at the University of Massachusetts. Link's research interests include marine resource–ecosystem modeling methodologies, marine food web topology, globally consistent patterns in ecosystem cumulative biomass distributions, and delineation of ecosystem overfishing thresholds. He is a fellow of the American Institute of Fishery Research Biologists and was previously a Frohlich fellow. Link holds executive leadership certificates from the Harvard Kennedy School and the Key Program at American University. He has received the Fisheries Society of the British Isles Medal for significant advances in fisheries science and a Department of Commerce Bronze Medal. Link received his B.S. in biology from Central Michigan University and a Ph.D. in biological sciences from Michigan Technological University.

Allison Miller is research portfolio senior manager at Schmidt Ocean Institute. She is responsible for managing and overseeing all research projects undertaken by Schmidt Ocean Institute, including grants and contracts, tracking cruise metrics, permitting, and ensuring scientists share the data they collect. Previously, Miller facilitated federal partnerships by managing the National Oceanographic Partnership Program, housed at the Consortium for Ocean Leadership. She serves as secretary of The Oceanography Society Council and serves on the external advisory committee of the University Corporation for Atmospheric Research Community Programs. Miller received a B.S. in marine science from Coastal Carolina University and an M.S. in oceanography from Florida State University.

S. Bradley Moran is dean of the College of Fisheries and Ocean Sciences at the University of Alaska Fairbanks. Prior to his appointment as dean, he served as acting director of the Obama Administration's National Ocean Council, assistant director for ocean sciences in the White House Office of Science and Technology Policy, program director in the Chemical Oceanography Program at the National Science Foundation, and professor of oceanography in the Graduate School of Oceanography at the University of Rhode Island. Moran's principal research interests focus on the application of uranium-series and artificial radionuclides as tracers of marine geochemical processes and fostering economic development partnerships in energy and environmental research, technology, policy, and education. He served as vice president of the international Scientific Committee on Oceanic Research; on the board of directors of the Alaska Ocean Observing System, the Alaska Sea Life Center, and the North Pacific Research Board; previously, he served as board chair and trustee of the Consortium for Ocean Leadership. Moran has served on the National Academies of Sciences, Engineering, and Medicine's Ocean Studies Board, and he is a member of the U.S. National Committee for the Decade of Ocean Science for Sustainable Development. Moran received a B.Sc. in chemistry from Concordia University and a Ph.D. in oceanography from Dalhousie University.

Richard W. (Rick) Murray is senior scientist (emeritus) in marine chemistry and geochemistry at Woods Hole Oceanographic Institution (WHOI) after serving at WHOI for five years as deputy director and vice president for science and engineering. Prior to this role, Murray was professor of Earth and environment at Boston University from 1992 to 2019, and he served as director for the Division of Ocean Sciences at the National Science Foundation from 2015 to 2018. Murray also served as co-chair for the Subcommittee on Ocean Science and Technology as part of the Office of Science and Technology Policy in the Executive Office of the President, during both the Obama and first Trump administrations. He has testified before the U.S. Congress and the Massachusetts State Legislature on issues relating to climate change and ocean observations. Murray's research interests are in marine geochemistry with an emphasis on sedimentary chemical records of climate change, volcanism, and tropical oceanographic processes, and in the chemistry of the subseafloor biosphere. He was involved in the advisory structure for scientific ocean drilling programs throughout his career and sailed on six different scientific drilling expeditions (127, 165, 175, 185, 329, and 346), including as co-chief scientist on Expedition 346. Murray is a former councilor of The Oceanography Society and a former member of the board of directors of the American Geophysical Union. He received a B.A. in geology from Hamilton College and a Ph.D. in geology and geophysics from the University of California, Berkeley. Murray also graduated from the Sea Education Association's program in Woods Hole, Massachusetts, and completed postdoctoral training at the University of Rhode Island's Graduate School of Oceanography.

Stephen R. Palumbi is Jane and Marshall Steel professor in marine sciences and senior fellow with the Woods Institute for the Environment at Stanford University. He is former director for Hopkins Marine Station at Stanford. Palumbi's research interests include the use of molecular genetics techniques to study evolution and change within marine populations. He has contributed to enhancing understanding of speciation patterns in open-ocean systems, providing insights for marine reserve design and refuges for thermally sensitive corals. Palumbi received the Peter Benchley Award for Excellence in Science and is an elected member of the National Academy of Sciences, fellow of the California Academy of Sciences, and Pew fellow in marine conservation. He has published three books on science for the general public, cofounded the microdocumentary series *Short Attention Span Science Theater*, and has appeared in numerous ocean documentaries. He received a B.A. in biology from Johns Hopkins University and a Ph.D. in zoology with a concentration in marine ecology from the University of Washington.

Ella (Josie) Quintrell recently retired as founding executive director of the Integrated Ocean Observing System (IOOS) Association and now serves as senior advisor to the organization. The IOOS Association works with the IOOS regional associations to design and operate coastal observing systems to collect, integrate, and produce information for users. Quintrell's research interests focus on operational observing issues related to harmful algal blooms, cloud computing, and coastal management. She previously served on the board of the Consortium for Ocean Leadership, the National Estuarine Research Reserve System Science Advisory Committee, and the National Marine Association of Marine Laboratories. Quintrell received a B.A. in biology from Colby College and an M.R.P. in environmental planning from Cornell University.

Yoshimi (Shimi) Rii is assistant specialist at Hawaiʻi Institute of Marine Biology and serves as research coordinator for the Heʻeia National Estuarine Research Reserve within the National Oceanic and Atmospheric Administration's Office for Coastal Management. Her expertise is in marine ecology with phytoplankton and nutrient dynamics in coastal and open-ocean environments. Rii utilizes multiple tools spanning geochemical tracers, biomarkers, and genomic approaches to examine metabolic activities and biodiversity of microbial eukaryotes. Her work encompasses research at the watershed to coastal ocean scale, examining the continuum and diversity of ecologically important species. She is passionate about expanding the impact of ocean science to multiple communities through the weaving of different disciplines and perspectives. Rii received a B.S. in marine biology and English from the University of California, Los Angeles, and an M.S. and Ph.D. in biological oceanography from the University of Hawaiʻi at Mānoa.

Kristen St. John is professor of geology at James Madison University. Her research focuses on marine sediment records of past climate change and on teaching and learning in the geosciences. As an active researcher in the International Ocean Discovery Program (IODP) and legacy programs, St. John participated as a marine sedimentologist for several expeditions; served as co-chief scientist for Expedition 403 in summer 2024 on the *JOIDES Resolution*; and is a co-proponent on an active IODP proposal for scientific drilling in the Arctic Ocean for consideration by the European Consortium for Ocean Research Drilling. She is the past-president of the American Geophysical Union Education Section and fellow of the Geological Society of America. St. John was editor-in-chief of the *Journal of Geoscience Education* from 2012 to 2017. She previously chaired the Workshop on Tipping Points, Cascading Impacts, and Interacting Risks in the Earth System and was a member of the Committee on Advancing a Systems Approach to Studying the Earth: A Strategy for the National Science Foundation. St. John served on the workshop steering committee and co-author of the report for the NEXT: Scientific Ocean Drilling Beyond 2023. Currently, she serves on the Polar Research Board. St. John received an M.S. and Ph.D. in geoscience from The Ohio State University.

Samuel K. (Kersey) Sturdivant is principal scientist at INSPIRE Environmental; adjunct assistant professor at Duke University; and cofounder of Oceanography for Everyone, an open-source effort to develop low-cost oceanographic hardware. His research interests center broadly around the effects of human disturbance on the seafloor and development of marine technology to enhance human understanding of the ocean. Sturdivant gave a TED Talk called "Visualize the Seafloor" and published a comprehensive step-by-step guide on how to get into graduate school in the sciences with Cambridge University Press. Sturdivant received a B.S. in environmental science from the University of Maryland Eastern Shore and a Ph.D. in marine science from the College of William & Mary's Virginia Institute of Marine Science.

Ajit Subramaniam is research professor at the Lamont-Doherty Earth Observatory (LDEO) of Columbia University. He has served as program director for the Marine Microbiology Initiative at the Gordon and

Betty Moore Foundation and as program director in the Biological Oceanography Program at the National Science Foundation while on leave from LDEO. Subramaniam is a microbial oceanographer with expertise in biogeochemical cycles, remote sensing, bio-optics, and phytoplankton physiology. His research interests focus on advancing the ability to observe the ocean and expand understanding on how the marine ecosystem works and can be managed. Subramaniam was awarded a Mercator fellowship by the University of Rostock and the Baltic Sea Research Institute, Germany, in 2017 and the Climate and Life fellowship at Lamont-Doherty Earth Observatory in 2021. He received a B.S. in physics from The American College in India, and an M.S. in marine environmental science and a Ph.D. in coastal oceanography from the State University of New York at Stony Brook.

Maya Tolstoy is Maggie Walker dean of the University of Washington College of the Environment. Prior to this role, Tolstoy was professor at Columbia University's Department of Earth and Environmental Sciences at Lamont-Doherty Earth Observatory, and she previously served as interim executive vice president and dean of the faculty of Arts and Sciences at Columbia. She is a marine geophysicist specializing in seafloor earthquakes and volcanoes. Over her more than 30-year career as a researcher, professor, and administrator, Tolstoy has dedicated herself to furthering understanding of the fundamental processes of the planet and broadening participation in academia. She was awarded the WINGS WorldQuest Sea Award honoring women in exploration and was a finalist for the National Aeronautics and Space Administration's 2009 astronaut selection. Tolstoy previously served on the National Academies of Sciences, Engineering, and Medicine's Board of Earth Sciences and Resources' Committee on Solid Earth Geophysics. She received a B.S. in geophysics from the University of Edinburgh and a Ph.D. from Scripps Institution of Oceanography at the University of California, San Diego.

Shannon Valley was most recently a climate advisor with Vistant, contracted to the United States Agency for International Development's Center for Climate Positive Development. Prior to this role, Valley served as a legislative liaison at the headquarters of the National Aeronautics and Space Administration (NASA) and as a policy assistant in the White House Domestic Policy Council. She was appointed to the 2020 presidential transition NASA agency review team. Valley's past research focused on Atlantic Meridional Overturning Circulation variability and its relation to abrupt and long-term climate change. She is a recipient of the NASA Individual Special Act Award, the NASA Exceptional Achievement Medal, and the National Science Foundation Graduate Research Fellowship. Valley received a B.A. in political science and international studies from Northwestern University and an M.S. and Ph.D. in Earth and atmospheric science from the Georgia Institute of Technology. She completed postdoctoral research in paleoceanography and coastal geochemistry, working with marsh sediment cores at Woods Hole Oceanographic Institution.

James Zachos is distinguished professor of Earth and planetary sciences and Ida Benson Lynn chair of ocean health at the University of California, Santa Cruz. His research focuses on resolving aspects of the ocean, climate, and carbon cycle dynamics of the last 65 million years, addressing issues ranging from the causes of extreme greenhouse warming and ocean acidification to the onset of Antarctic glaciation. Zachos's past research included participation in the scientific ocean drilling program. He is a member of both the American Academy of Arts and Sciences and the National Academy of Sciences and is a fellow of the American Geophysical Union (AGU), the Geological Society of America, and the American Association for the Advancement of Science. Zachos is also a recipient of the AGU Emiliani Award, the European Geophysical Union Milutin Milankovic Medal, and the BBVA Frontiers of Knowledge Award. He received a B.S. in geology and economics from the State University of New York at Oneonta, an M.S. in geology from the University of South Carolina, and a Ph.D. from the Graduate School of Oceanography at the University of Rhode Island. He completed a postdoctoral fellowship at the University of Michigan.

Appendix B
Committee Meeting Topics

COMMITTEE MEETING #1
June 21–22, 2023
Washington, DC

TOPIC: Sponsor Briefing and Discussion
Jim McManus, Director, Division of Ocean Sciences, National Science Foundation

TOPIC: Overview of NSF's Directorate on Technology, Innovation, and Partnerships
Allen Walker, Senior Advisor, Technology, Innovation, and Partnerships, National Science Foundation

COMMITTEE MEETING #2
August 2–3, 2023
Washington, DC

TOPIC: Overview of International Ocean Discovery Program (IODP)
Mitch Malone, Texas A&M University

TOPIC: Briefing from July IODP Town Hall
Jim McManus, Division Director, Division of Ocean Sciences, National Science Foundation

TOPIC: Briefing on the 2050 Framework of Scientific Ocean Drilling
Anthony Koppers, Oregon State University

TOPIC: The Future of Scientific Ocean Drilling
Anthony Koppers, Oregon State University
Adriane Lam, Binghamton University
Patrick Fulton, Cornell University
Kathie Marsaglia, California State University, Northridge

TOPIC: A Future Vision from Ocean Drilling Legacy Assets Projects (LEAPs) Report
Larry Krissek, Ohio State University

TOPIC: Benefits to Overlapping, Cross-disciplinary Ocean and Pale-Ocean Research Priorities
Daniel Sigman, Princeton University

TOPIC: Future Perspectives on Scientific Ocean Drilling
Masako Tominaga, Woods Hole Oceanographic Institution
Chijun Sun, National Center for Atmospheric Research
Adriane Lam, Binghamton University
Chris Lowery, University of Texas
Allyson Tessin, Kent State University
Jason Sylvan, Texas A&M University
Jessica Labonté, Texas A&M University, Galveston

Brandi Kiel Reese, University of South Alabama
Patrick Fulton, Cornell University

TOPIC: Briefing from Science Mission Requirements
Becky Robinson, University of Rhode Island

TOPIC: Panel Discussion on New Options in Infrastructure
Sean Gulick, University of Texas
Maureen Walczak, Oregon State University
Becky Robinson, University of Rhode Island
Carl Brenner, U.S. Science Support Program
Rick Murray, Committee Member

COMMITTEE MEETING #3
October 24–25, 2023
Portland, OR

TOPIC: Ocean Observing Infrastructure and Innovation
Deborah Kelley, University of Washington
Ed Dever, Oregon State University
Jim Edson, Woods Hole Oceanographic Institute

TOPIC: Challenges and Opportunities in Physical Oceanography
LuAnne Thompson, University of Washington
Craig Lee, University of Washington
Melanie Fewings, Oregon State University
Jack Barth, Oregon State University/Northwest Association of Networked Ocean Observing
    Systems (NANOOS)
Joe Schumacker, NANOOS

TOPIC: Artificial Intelligence/Machine Learning
Patrick Heimbach, University of Texas
Prasanna Sattigeri, IBM
Kanna Rajan, Rand

TOPIC: Ocean Solutions: Co-Design and Co-Development
Rosie Alcgado, University of Hawai'i
Kirsten Oleson, Pacific Research on Island Solutions for Adaptation (RISA)
Katie Arkema, Pacific Northwest National Lab
Charlie Plybon, Surfrider Foundation

TOPIC: Research Priorities for National Science Foundation Division of Ocean Sciences (OCE): Ocean
    Acidification/Deoxygenation/Harmful Algal Blooms (HABs)
Maria Kavanaugh, Oregon State University
Alexis Valauri-Orton, The Ocean Foundation

TOPIC: Research Priorities for OCE: Marine Carbon Dioxide Removal (CDR)
David Koweek, OceanVisions
Julie Pullen, Propeller Ventures

TOPIC: Research Priorities for OCE: Critical Minerals
   Beth Orcutt, Bigelow Laboratory for Ocean Sciences
   Amy Gartman, United States Geological Survey

COMMITTEE MEETING: NOVEMBER VIRTUAL
November 20, 2023

TOPIC: Marine Long-Term Ecological Research (LTER) Panel and Discussion
   Russ Hopcroft, Northern Gulf of Alaska LTER, University of Alaska Fairbanks
   Katherine Barbeau, California Current Ecosystem LTER, University of California,
    San Diego
   Mark Ohman, California Current Ecosystem LTER, University of California, San Diego
   Merryl Alber, Georgia Coastal Ecosystem LTER, University of Georgia

COMMITTEE MEETING: JANUARY VIRTUAL
January 31, 2024

TOPIC: Subduction Zone Geohazards
   Emily Brodsky, SZ4D, University of California, Santa Cruz
   Doug Wiens, SZ4D, Washington University
   Diego Melgar, CRESCENT, University of Oregon
   William Wilcock, COSZO, University of Washington

COMMITTEE MEETING #4
February 15–16, 2024
Gulfport, MS

TOPIC: Impact of the 2015–2025 Decadal Survey
   Jim McManus, National Science Foundation
   Rick Murray, Committee Member

TOPIC: Status and Future of the Academic Research Fleet
   Rose Dufour, National Science Foundation
   Deborah Bronk, Bigelow Laboratory for Ocean Sciences/University-National Oceanographic
    Laboratory System (UNOLS) Chair
   Doug Russell, UNOLS Executive Secretary
   Robert Shearman, Office of Naval Research
   Robert Sparrock, Office of Naval Research

TOPIC: Public–Private Partnerships in Ocean Sciences
   Henry Jones, University of Southern Mississippi
   Karen Grissom, National Oceanic and Atmospheric Administration (NOAA) National Centers
    for Environmental Information (NCEI)
   Tim Boyer, NOAA NCEI
   Almesha Campbell, Jackson State University

TOPIC: A Future Equitable Ocean Sciences
   Corey Garza, University of Washington

Christine Yifeng Chen, Lawrence Livermore National Laboratory
Robert Twilley, Louisiana State University
Brandon Jones, National Science Foundation
Aradhna Tripati, University of California, Los Angeles

TOPIC: Opportunities for AI/ML to Advance Ocean Sciences
Warren Wood, U.S. Naval Research Laboratory
Peter Gerstoft, Scripps Institution of Oceanography
Heidi Sosik, Woods Hole Oceanographic Institution

TOPIC: Ocean Life, Part 1—Overview of Biodiversity Issues
Gabrielle Canonico, National Oceanic and Atmospheric Administration
Emmett Duffy, Smithsonian Institution
Joey Bernhardt, University of Guelph

TOPIC: Ocean Life, Part 1—Response to Climate Change
Malin Pinsky, University of California, Santa Cruz
Steven Murawski, University of South Florida
Sarah Davies, Boston University

TOPIC: Ocean Life, Part 1—New Tools in Discovering Biodiversity: eDNA
Paul Barber, University of California, Los Angeles
Zachary Gold, National Oceanic and Atmospheric Administration

COMMITTEE MEETING: MARCH VIRTUAL
March 26, 2024

TOPIC: The Oceans and Climate: Sustained Global Observing Systems (OneArgo and GO-SHIP)
Lynne Talley, Scripps Institute of Oceanography
Ken Johnson, Monterey Bay Aquarium Research Institute
Susan Wijffels, Woods Hols Oceanographic Institute

COMMITTEE MEETING: APRIL VIRTUAL
April 15, 2024

TOPIC: Information Gathering Session with Futurist, Duane Elgin
Duane Elgin, Author of *Choosing Earth*

COMMITTEE MEETING: APRIL VIRTUAL
April 24, 2024

TOPIC: A Discussion with NSF's Division of Geosciences
Alex Isern, Assistant Director of Geosciences, National Science Foundation

TOPIC: SMART Cables
Bruce Howe, University of Hawai'i at Mānoa
Zhongwen Zhan, California Institute of Technology

COMMITTEE MEETING #5
May 21–22, 2024
Narragansett, RI

TOPIC: Research Priorities in Marine Geology and Geophysics
Ross Parnell-Turner, Scripps Institution of Oceanography
Jessica Warren, University of Delaware
Frieder Klein, Woods Hole Oceanographic Institution
Jackie Caplan-Auerbach, Western Washington University

TOPIC: Ocean Life, Part II
Kristy Kroeker, University of California, Santa Cruz
Sonya Dyhrman, Lamont-Doherty Earth Observatory
Mike Stukel, Florida State University
Erik Cordes, Temple University
David Hutchins, University of Southern California

TOPIC: Urban Seas and Coastal Oceans Research Priorities
John Delaney, University of Washington
Robert Twilley, Louisiana State University
Holly Greening, Tampa Bay Estuary Program
Jeremy Testa, University of Maryland Center for Environmental Science (UMCES) Chesapeake
     Biological Laboratory
Sarah Giddings, Scripps Institution of Oceanography
Robert Sterner, University of Minnesota Duluth
Margaret McManus, University of Hawai'i

COMMITTEE MEETING: JUNE VIRTUAL
June 25, 2024

TOPIC: Coupled Ocean–Atmosphere Modeling and Observations
Galen McKinley, Columbia University
Eric Chassignet, Florida State University
Shuyi Chen, University of Washington
Sarah Gille, Scripps Institution of Oceanography